PLANT SCIENCE RESEARCH AND PRACTICES

NECTAR

PRODUCTION, CHEMICAL COMPOSITION AND BENEFITS TO ANIMALS AND PLANTS

Plant Science Research and Practices

Additional books in this series can be found on Nova's website under the Series tab.

Additional e-books in this series can be found on Nova's website under the e-book tab.

PLANT SCIENCE RESEARCH AND PRACTICES

NECTAR

PRODUCTION, CHEMICAL COMPOSITION AND BENEFITS TO ANIMALS AND PLANTS

ROBIN L. PECK
EDITOR

nova publishers
New York

Library of Congress Cataloging-in-Publication Data

Nectar : production, chemical composition and benefits to animals and plants / editor: Robin L. Peck.
pages cm. -- (Plant science research and practices)
Includes index.
ISBN 978-1-63463-650-6 (hardcover)
1. Honey plants. 2. Nectar. I. Peck, Robin L., editor. II. Series: Plant science research and practices.
SF535.N43 2015
638'.13--dc23

2014045014

Published by Nova Science Publishers, Inc. † New York

Contents

PREFACE

This book discusses several benefits of plant nectar. It also discusses the production and chemical composition of nectar.

Chapter 1 - Floral nectar is a sweet, aqueous secretion that mainly consists of sugars and to a lesser extent amino acids. It is offered by plants to entice pollinators and to ensure pollination and thus sexual reproduction. Although floral nectar is initially sterile, some yeasts and bacterial species frequently inhabit floral nectar, once they have been dispersed into the nectar via the air or flower-visiting insects such as pollinators. Identification of microorganisms in floral nectar from broad-scale surveys has shown that nectar yeasts are widespread and occur in a wide range of habitats. No such surveys are currently available for bacteria, but the few available studies suggest that bacteria may also be widespread in nectar. However, the high sugar content (sometimes till 50% w/w) and consequentially low water activity (typically between 0.90 and 0.99 a_w) that characterize the nectar of many plant species limit the number of microbial species that are capable of surviving and proliferating in the harsh environment of nectar, leading to species-poor communities that are often dominated by one or few culturable species per nectar sample. Other factors that determine nectar communities are nitrogen availability, pH, sugar composition, temperature and the presence of secondary or antimicrobial compounds. The presence of microbes in nectar has been shown to decrease sugar concentrations and alter sugar ratios. Besides, the metabolic activity of nectar-inhabiting microorganisms also affects other floral attractive traits, including temperature, scent and amino acid content. In the few cases that have been examined, changes in nectar chemical properties impacted on pollinator behaviour and as a result on plant reproductive success. A broader investigation of the impact of bacteria and yeasts on the chemical and physiological properties of nectar, coupled with behavioural assays of pollinators, are needed to translate microbial presence in floral nectar to plant population dynamics.

Chapter 2 - Onion is the second most cultivated vegetable crop in the world. Two types of onion varieties are usually grown around the world; open pollinated (OP) and first generation (F1) hybrids. Honey bees are the main pollinators of onion seed production, but owing to low attractiveness of flowers, pollination is often inadequate. Pollination problems result in low seed yields. Field observations indicate that F1 hybrid seed yields are much lower than open-pollinated varieties seed yields, with a decrease of up to 60%. Nectar is the most important floral reward offered by plants. Floral nectar composition has been studied as the potential cause of bee preferences in many species, among them onion. Nectar usually does not repel bees, but some nectar may be less attractive than nectar of competing flowers. Honeybees have distaste for onion flowers and avoid visiting them. Frequently, onion nectar potassium concentration and nectar sucrose concentration have been mentioned as potential causes of bee unattractiveness. Also, minor components of nectar such as phenols, amino acids, proteins, organosulfur compounds and alkaloids might directly affect bee foraging. There are differences in the chemical composition of nectar and floral morphology between open pollinated and F1 Hybrids and also within F1 Hybrids. These differences are correlated with the number of visits and seed yield. Therefore, it would be possible to select indirectly the most promising seed production lines analyzing nectar or floral morphological characters.

Chapter 3 - In plant-ant mutualism, plants often provide foods, such as extrafloral nectar and food bodies, to ants. In return, the visiting ants protect the plants against herbivores. However, these plants also possess various direct defenses, including physical defense via trichomes and tough leaves, chemical defense using secondary metabolites. Why have plants evolved defensive mutualisms with ants even when possessing these direct defenses? To answer this question, the authors must clarify the costs and benefits of each defense trait, as well as its role in the overall defense strategy of the plant. Previous studies, however, have mostly dealt with the costs and benefits of particular defenses. Here, the authors review recent advances in the study of plant defense strategies involving the indirect defense provided by extrafloral nectar production and other defenses. First, they explain the benefits of each defense trait, with special reference to its efficacy against herbivores. Second, they review the costs of each defense trait. Next, the authors focus on differences in the use of each defense trait in response to habitat conditions and the value of plant tissue.

Finally, the authors highlight the major gaps in their current knowledge and outline future research directions.

Chapter 4 - To search for factors governing butterfly diversity patterns, the authors studied adult butterfly communities and their diet resources in and around the Aokigahara primeval woodland of Mount Fuji, central Japan. In case study 1, they established two study sites in the woodland interior, one at the edge of the woodland, and two in the adjacent semi-natural grassland. The authors monitored adult butterflies and their food items from May to October in 2005 using a transect method. The total number of butterfly species recorded was greatest in the woodland edge site (41 spp.), followed by those in semi-natural grassland sites (39 and 38 spp.). Fewer species were recorded in the woodland interior sites (28 and 20 spp.). This finding supports the commonly held view that butterfly diversity is highest in intermediate successional stages. They also recorded the 38 food items used by adult butterflies. Most were Spermatophyte plants with nectar-bearing flowers. Adult butterflies particularly favored the flowers of perennial herbaceous plants. The total number of butterfly species in each site was significantly positively correlated with the number of herbaceous plant species present. Therefore, the high butterfly species richness in the mid-successional woodland edge and semi-natural grassland habitats appears to be due to the variety of favored food plants in these sites. Seven species mentioned in the Ministry of the Environment Red List were observed during this study. Their populations were predominantly concentrated in the semi-natural grassland sites, suggesting that the preservation of these habitats is important for conserving these endangered butterfly species. In case study 2, the authors analyzed the relationships between the diversities of vegetation, adult nectar plants, and butterflies in sites in and around the woodland different from those of the case study 1. As well as the results of the case study 1, the nectar resource utilization by adult butterflies deviated extremely to herbaceous plants, especially to perennials, compared to woody species, although most of the study area was in and near the woodland. There were more nectar plant species in sites with more plant species richness. Among the butterfly community indices analyzed, the strongest correlation was detected between butterfly species richness and nectar plant species richness in each site. Another close correlation was detected between the species richness of nectar plants and herbaceous plants in each site. In conclusion, nectar plant species richness is a highly important factor supporting adult butterfly species richness in and around woodland areas. Additionally, the maintenance of herbaceous plant species richness in a butterfly habitat, which leads to the maintenance of nectar plant species richness, is very important for conservation of butterfly diversity even in and around woodland landscape of temperate region.

Chapter 5 - Some flower morphophysiological and pollinator-luring features as well as co-adaptive traits, including nectar production, nectaries and pollen of five Lamiaceae species, namely *Hyssopus officinalis, Lavandula officinalis, Melissa officinalis, Salvia officinalis* and *Salvia sclarea*, were studied, with reference to the pollinator visits, especially considering honey bee forage preferences. Although these species are cultivated in plantation for medicinal purposes, at the same time they represent a significant source of pollen and nectar for honey production, therefore the research aimed to estimate their melliferous potential, attractiveness and significance for bee pasture.

The pollen grains are isopolar, radially symmetric, predominantly hexacolpate in all species, but often mixed together with tetracolpate and octacolpate in *S. sclarea*. They are mostly medium-sized to large. The smallest grain was found in *L. officinalis* and the largest one in *Salvia sclarea*. The exine sculpturing is bireticulate (*H. officinalis, M. officinalis, S. officinalis*), reticulate (to foveolate) (*S. sclarea*) and foveolate (to perforate) (*L. officinalis*). The shape of pollen grains in equatorial view varies among different species, and within the same species. In the majority of studied species, shape is most frequently prolate (*L. officinalis* - type a, *M. officinalis, S. officinalis* and *S. sclarea* - Class I), except in *H. officinalis* and *S. sclarea* - Class II, in which they are sub-prolate and *L. officinalis* - type b in which they are oblate-spheroidal.

Floral nectaries are located on the receptacle, surrounding the ovary base in the shape of an asymmetrical annular disc with four lobes. Unevenly distributed modified stomata were observed on the upper part or the entire surface of the functional, and generally, the most developed nectary lobe facing the lower corolla lip. The nectariferous tissue is composed of the uniseriate epidermis overlying several layers of polygonal cells forming subepidermal secretory tissue. Epidermal cells are lined with very thin cuticle. Xylem elements alone, separating from the vasculature of the flower receptacle, were noticeable nearby or inside the nectariferous tissue.

The highest total daily and summary diurnal nectar amount per flower, as well as the nectar secretion rate were found in *S. sclarea,* and the lowest in the M. *officinalis* respectively. Pattern of diurnal dynamics of nectar secretion with one secretion peak was observed: in the morning, in both *Salvia* species, *H. officinalis* and *L. officinalis,* and early in the afternoon in *M officinalis.*

Pollinator observation indicated the visitation of seven species of Hymenoptera, three species belonging to Diptera and two species of daily-active Lepidoptera. The honey bee (*Apis mellifera*) was the most abundant pollinator. The major elements of attractiveness were flower scent and colour, followed by the shape and size of the corolla and to a lesser extent the nectar amount.

According to all analyzed flower characteristics and observation of pollinator behaviour and visits, the most attractive plants were *H. officinalis, S. officinalis* and *L. officinalis*, whereas *M. officinalis* was the least attractive. *S. sclarea* was completely unattractive to all recorded pollinators, except *Xylocopa violacea*.

In: Nectar
Editor: Robin L. Peck

ISBN: 978-1-63463-650-6

Chapter 1

IMPACT OF MICROORGANISMS ON NECTAR CHEMISTRY, POLLINATOR ATTRACTION AND PLANT FITNESS

María I. Pozo[1]*, Bart Lievens[2] and Hans Jacquemyn[1]
[1]KU Leuven, Biology Department,
Plant Population and Conservation Biology, Heverlee, Belgium
[2]KU Leuven, Department of Microbial and Molecular Systems (M^2S),
Laboratory for Process Microbial Ecology and Bioinspirational
Management (PME&BIM), Campus De Nayer,
Sint-Katelijne-Waver, Belgium

ABSTRACT

Floral nectar is a sweet, aqueous secretion that mainly consists of sugars and to a lesser extent amino acids. It is offered by plants to entice pollinators and to ensure pollination and thus sexual reproduction. Although floral nectar is initially sterile, some yeasts and bacterial species frequently inhabit floral nectar, once they have been dispersed into the nectar via the air or flower-visiting insects such as pollinators. Identification of microorganisms in floral nectar from broad-scale surveys has shown that nectar yeasts are widespread and occur in a wide range of habitats. No such surveys are currently available for bacteria, but the few available studies suggest that bacteria may also be widespread in nectar.

* Corresponding author: maribel.pozoromero@bio.kuleuven.be.

However, the high sugar content (sometimes till 50% w/w) and consequentially low water activity (typically between 0.90 and 0.99 a_w) that characterize the nectar of many plant species limit the number of microbial species that are capable of surviving and proliferating in the harsh environment of nectar, leading to species-poor communities that are often dominated by one or few culturable species per nectar sample. Other factors that determine nectar communities are nitrogen availability, pH, sugar composition, temperature and the presence of secondary or antimicrobial compounds. The presence of microbes in nectar has been shown to decrease sugar concentrations and alter sugar ratios. Besides, the metabolic activity of nectar-inhabiting microorganisms also affects other floral attractive traits, including temperature, scent and amino acid content. In the few cases that have been examined, changes in nectar chemical properties impacted on pollinator behaviour and as a result on plant reproductive success. A broader investigation of the impact of bacteria and yeasts on the chemical and physiological properties of nectar, coupled with behavioural assays of pollinators, are needed to translate microbial presence in floral nectar to plant population dynamics.

"*All microbial ecology is chemical ecology*"- Mark Hay, cited in Raguso (2014, J Chem Ecol 40:412–413)

INTRODUCTION

About 90% of flowering plant species rely on pollinators for reproduction and maintenance of genetic diversity (Ollerton et al., 2011). Floral nectar is by far the most common means of plants to entice pollinators and to ensure reproductive success (Nicholson & Thornburg, 2007; Heil, 2011). Nectar composition determines the spectrum of flower-visiting animals and nectar consumers, including spiders, holometabolous insects (including the main three orders Hymenoptera, Diptera and Lepidoptera), birds, reptiles, and mammals (Nicolson, 2007), many of which act as pollinators. Plant products, such as nectar, that are secreted to attract animal pollinators should try to match their nutritional requirements (Nicolson, 2007). Baker and Baker (1983) demonstrated convergence in nectar sugar composition between plants with the same visitors or sharing flower traits that evolved in response to natural selection imposed by pollinators (this phenomenon is known as "pollination syndrome"). Hummingbirds, butterflies, moths and long-tongued bees usually prefer sucrose-rich nectar, whereas short-tongued bees and flies prefer hexose-rich nectar (Baker & Baker, 1983). The preference for hexose-rich nectar has been related to some physiological constraints in nectar feeders, as in the specific case of some nectarivorous birds and ants that lack the enzyme invertase and are not able to break down sucrose (Napier et al., 2013).

Although the characteristics of nectar tend to be similar between plants that exhibit the same pollination syndrome (Fenster et al., 2004), the plant itself (phylogenetic component) also bears a strong imprint on sugar composition (Ornelas et al., 2007). In some plant groups, sugar composition is highly conservative and reflects taxonomic relationships rather than pollinator types (Elisens & Freeman, 1988; Van Wyk et al., 1993; Galetto et al., 1998; Nicolson & Van Wyk, 1998), indicating that plant phylogeny can be a stronger determinant of nectar composition than visitor guild (Galetto and Bernardello, 2004). However, with the advent of new technological methods that allow more detailed quantification of sugar composition, it has become clear that nectar composition may show greater intraspecific heterogeneity than previously thought and is affected by other factors such as climate or geographical distribution (Galetto & Bernardello, 2003; Petanidou, 2005; Nepi et al., 2010). Sugar composition has also been shown to vary between the two sexual phases in hermaphroditic flowers (Langenberger & Davis, 2002; Canto et al., 2011), according to plant phenology (Herrera et al., 2006), between flowers of different plants (Herrera et al., 2006) and between populations with different floral visitors (Chalcoff et al., 2006). In the most extreme case, variation in sugar composition has even been observed between nectaries within the same flower (Herrera et al., 2006).

While some of this variation can be attributed to the plant itself, recent research has suggested that large differences in nectar chemistry at the within individual plant level are most likely the result of microbial activity (Herrera et al., 2006; Canto et al., 2007). Indeed, nectar microbes are often vectored by air and by flower-visiting animals into the floral nectar (Canto et al., 2008). Nectar is colonized by a range of microorganisms, including several yeasts and bacteria (Lievens et al., 2014), which may become extremely abundant in the nectar of many plants, reaching densities of over 10^6 cells/mm³ for yeasts (Brysch-Herzberg, 2004; de Vega et al., 2009; Herrera et al., 2009) and over 10^2 cells/mm³ for bacteria (Fridman et al., 2011). Yeast metabolism causes a density-dependent effect on nectar chemistry, such as a decline in total sugar concentration (Herrera et al., 2008, de Vega et al., 2009), changes in sugar composition and increase in nectar temperature (Herrera & Pozo, 2010). Additionally, flower odor may be modified by the emission of yeast volatiles (Golonka et al., 2014). While direct evidence is not yet available for nectar bacteria, similar trends may be expected as those caused by yeasts (Vannette et al., 2013). Due to changes in nectar chemistry, the presence of microorganisms in floral nectar may have broad ecological implications through its effects on pollinators, and several studies have mentioned the relevance of nectarivorous microbes in plant-pollinators relationships (e.g., Kevan et al., 1988; Eisikowitch et al., 1990; Canto et al., 2008; de Vega et al., 2009; Herrera et al., 2009; Pozo et al., 2009; Herrera et al., 2013; Vannette et al., 2013; Good et al., 2014). The strength of this tripartite interaction, however, depends on the extent to which nectar microbes impact on pollinator foraging behaviour.

In this chapter, we provide an overview of the potential impact of microorganisms on nectar chemistry, pollinator attraction and plant fitness. We first discuss the main features of floral nectar. We then give a comprehensive overview of the main yeast and bacterial species occurring in nectar and highlight what is known of their ecological characteristics. Next, we investigate how microbial communities establish in floral nectar and how microbes are able to cope with varying nectar conditions when they are transferred from one plant species to the next. We then investigate how microbes affect the chemical and physical properties of nectar and as a result impact on pollinator behaviour and ultimately plant fitness. Finally, we highlight avenues for future research.

General Features of Floral Nectar

Nectar is a sugar-rich solution that mainly consists of mono- and disaccharides (Percival, 1961; Nicolson, 2007) and to a much lower extent of amino acids and other compounds such as lipids, minerals and vitamins (Carter et al., 2006). Most nectars are largely dominated by sucrose (de la Barrera & Nobel, 2004; Petanidou, 2005; Herrera et al., 2006; Canto et al., 2007), which is synthesized in the nectary parenchyma and subsequently secreted into the extracellular space via transporters like SWEET9 (Lin et al., 2014). In some plant species, however, sucrose can be hydrolyzed into its monomers by an apoplasmic invertase, yielding nectars composed by a mixture of sucrose, glucose and fructose (Nicolson, 2007). Several nectars also contain other carbohydrates such as melezitose, maltose, raffinose, and melibiose (Percival, 1961, Baker and Baker, 1983), though usually in minor concentrations (Baker & Baker, 1977, 1985, Lenaerts et al., unp. data). Nectar sugar composition is determined by intrinsic plant characteristics and bears a strong phylogenetic imprint (Nicolson & Thornburg, 2007; Nocentini et al., 2013). In contrast, nectar sugar concentration is highly determined by extrinsic features unrelated to the plant, including several environmental variables such as air humidity, soil moisture content, temperature, etc. (Corbet et al., 1979). Total sugar concentration varies significantly between individuals within species and between plant species, ranging from less than 10% to 66.5% (w/w) as average concentrations in nectar from *Aloe* and *Carum*, respectively (Nicolson & Thornburg, 2007).

Although sugar clearly dominates the solutes in nectar and represents the major energy source for pollinators, amino acids, which are 100 to 1000 times less concentrated than sugar, can also significantly affect the attractiveness of nectar (Herrera, 1989; Heil, 2013, Nepi, 2013). Nectar amino acid composition and its concentration have been related to some plant features, including flower age or flowering season (Petanidou et al., 1996). Additionally, it has also been related to pollinator attraction (Baker & Baker, 1986; Petanidou et al., 2006). In particular, amino acids are supposed to provide taste to nectar (Gardener & Gillman, 2002), and they affect insect behaviour by stimulating chemosensory receptors. For example, proline and phenylalanine would have a phagostimulant – feeding is induced by means of this substance – effect (Nicolson, 2007). While nectar-feeding birds and bats can also gain nitrogen from other sources, many adult insects just rely on pollen and nectar as sources to get all the essential amino acids (Nicolson, 2007). In addition, volatile organic compounds (VOCs) have been related to pollinator attraction, and scented petals have been known for centuries (Raguso, 2008). By contrast, nectar odours were only since a decade considered as a relevant signal to attract pollinators (Raguso, 2004). The origin of nectar scent has been usually linked to volatiles released by the petals that are absorbed and rereleased by the nectar (Raguso, 2008).

MICROORGANISMS INHABITING FLORAL NECTAR

Although it has already been known since the nineteenth century that microbes are common inhabitants of floral nectars (Boutroux, 1884; Schuster & Ulehla, 1913; Grüss, 1917; Schoelhorn, 1919; Nadson & Krassilnikov, 1927; Capriotti, 1953; Vörös-Felkai, 1957; Sandhu & Waraich, 1985; Lachance et al., 2001; Brysch-Herzberg, 2004), only recently the microbial community structure in nectar and its ecological impact have been explored in more detail (Table 1 and 2; Vannette et al., 2013, Herrera et al., 2013, Good et al., 2014).

Table 1. Overview of culturable yeast species found in floral nectar

Phylum	Class	Family	Genus	Species	Reference
Ascomycota	Dothideomycetes	Dothioraceae	*Aureobasidium*	*pullulans*	Álvarez-Pérez and Herrera, 2013
					de Vega and Herrera, 2012
					Golonka and Vilgalys, 2013
					Pozo et al., 2011; 2012
Ascomycota	Eurotiomycetes	Myxotrichaceae	*Geomyces*	*pannorum*	Jacquemyn et al., 2013a
Ascomycota	Saccharomycetes	Debaryomycetaceae	*Debaryomyces*	*hansenii*	Brysch-Herzberg, 2004
Ascomycota	Saccharomycetes	Debaryomycetaceae	*Debaryomyces*	*castellii*	Mushtaq et al., 2007; 2008
Ascomycota	Saccharomycetes	Debaryomycetaceae	*Debaryomyces*	*maramus*	de Vega and Herrera, 2012
Ascomycota	Saccharomycetes	Debaryomycetaceae	*Debaryomyces*	*polymorphus*	de Vega and Herrera, 2012
Ascomycota	Saccharomycetes	Debaryomycetaceae	*Debaryomyces*	*vanrijii*	Mushtaq et al., 2008
Ascomycota	Saccharomycetes	Incertae sedis	*Starmerella*	*bombicola*	Álvarez-Pérez and Herrera, 2013
					Golonka and Vilgalys, 2013
Ascomycota	Saccharomycetes	Lipomycetaceae	*Lipomyces*	*starkeyi*	Mushtaq et al., 2008
Ascomycota	Saccharomycetes	Metschnikowiaceae	*Metschnikowia*	*caudata*	de Vega et al., 2014
Ascomycota	Saccharomycetes	Metschnikowiaceae	*Metschnikowia*	*drakensbergensis*	de Vega et al., 2014
Ascomycota	Saccharomycetes	Metschnikowiaceae	*Metschnikowia*	*koreensis*	Brysch-Herzberg, 2004
					Canto and Herrera, 2012
					Golonka and Vilgalys, 2013

Phylum	Class	Family	Genus	Species	Reference
Ascomycota	Saccharomycetes	Metschnikowiaceae	*Metschnikowia*	*gruessi*	Álvarez-Pérez and Herrera, 2013
					Brysch-Herzberg, 2004
					de Vega and Herrera, 2012
					Pozo et al., 2011; 2012
Ascomycota	Saccharomycetes	Metschnikowiaceae	*Metschnikowia*	*reukaufii*	Álvarez-Pérez and Herrera, 2013
					Belisle et al., 2012
					Brysch-Herzberg, 2004
					de Vega and Herrera, 2012
					Golonka and Vilgalys, 2013
					Jacquemyn et al., 2013a
					Manson et al., 2007
					Pozo et al., 2011
					Vanette et al., 2013
Ascomycota	Saccharomycetes	Metschnikowiaceae	*Metschnikowia*	*proteae*	de Vega et al., 2012
Ascomycota	Saccharomycetes	Metschnikowiaceae	*Metschnikowia*	*vanudenii*	Golonka and Vilgalys, 2013
Ascomycota	Saccharomycetes	Pichiaceae	*Pichia*	*angusta*	Mushtaq et al., 2007; 2008
Ascomycota	Saccharomycetes	Pichiaceae	*Pichia*	*fabiantii*	Mushtaq et al., 2008
			Pichia	*guilliermondii*	Golonka and Vilgalys, 2013
Ascomycota	Saccharomycetes	Pichiaceae	*Pichia*	*jadinii*	Mushtaq et al., 2007; 2008
Ascomycota	Saccharomycetes	Pichiaceae	*Pichia*	*lynferdii*	Mushtaq et al., 2007; 2008
Ascomycota	Saccharomycetes	Pichiaceae	*Pichia*	*methanolica*	Mushtaq et al., 2007

Table 1. (Continued)

Phylum	Class	Family	Genus	Species	Reference
Ascomycota	Saccharomycetes	Pichiaceae	*Pichia*	*ofunaensis*	Mushtaq et al., 2007
Ascomycota	Saccharomycetes	Saccharomycetaceae	*Candida*	*bombi*	Golonka and Vilgalys, 2013
					Jacquemyn et al., 2013a
Ascomycota	Saccharomycetes	Saccharomycetaceae	*Candida*	*cfr. lactis-condensi*	Jacquemyn et al., 2013a
Ascomycota	Saccharomycetes	Saccharomycetaceae	*Candida*	*etchellsii*	Canto and Herrera, 2012
Ascomycota	Saccharomycetes	Saccharomycetaceae	*Candida*	*gropengiesseri*	Mushtaq et al., 2008
Ascomycota	Saccharomycetes	Saccharomycetaceae	*Candida*	*ipomoeae*	Canto and Herrera, 2012
Ascomycota	Saccharomycetes	Saccharomycetaceae	*Candida*	*magnoliae*	Álvarez-Pérez and Herrera, 2013
					Mushtaq et al., 2007
Ascomycota	Saccharomycetes	Saccharomycetaceae	*Candida*	*melibiosica*	de Vega and Herrera, 2012
Ascomycota	Saccharomycetes	Saccharomycetaceae	*Candida*	*railenensis*	de Vega and Herrera, 2012
					Serjeant et al., 2008
Ascomycota	Saccharomycetes	Saccharomycetaceae	*Candida*	*rancensis*	Brysch-Herzberg, 2004
					Manson et al., 2007
Ascomycota	Saccharomycetes	Saccharomycetaceae	*Candida*	*rhagii*	Mushtaq et al., 2008
Ascomycota	Saccharomycetes	Saccharomycetaceae	*Candida*	*sorbosivorans*	Canto and Herrera, 2012
Ascomycota	Saccharomycetes	Saccharomycetaceae	*Candida*	*succiphila*	Mushtaq et al., 2007; 2008
Ascomycota	Saccharomycetes	Saccharomycetaceae	*Candida*	*valdiviana*	Mushtaq et al., 2008
Ascomycota	Saccharomycetes	Saccharomycetaceae	*Candida*	*xestobii*	Mushtaq et al., 2008

Phylum	Class	Family	Genus	Species	Reference
Ascomycota	Saccharomycetes	Saccharomycetaceae	*Candida*	*gelsemii*	Manson et al., 2007
Ascomycota	Saccharomycetes	Saccharomycodaceae	*Hanseniaspora*	*uvarum*	Jacquemyn et al., 2013b
Ascomycota	Saccharomycetes	Saccharomycetaceae	*Issatchenkia*	*occidentalis*	Mushtaq et al., 2008
Ascomycota	Saccharomycetes	Saccharomycetaceae	*Saccharomyces*	*cerevisiae*	Dandu and Dhabe, 2011
Ascomycota	Saccharomycetes	Saccharomycetaceae	*Saccharomyces*	*kluyveri*	Mushtaq et al., 2007
Ascomycota	Saccharomycetes	Saccharomycetaceae	*Williopsis*	*californica*	Mushtaq et al., 2008
Ascomycota	Saccharomycetes	Saccharomycetaceae	*Williopsis*	*pratensis*	Mushtaq et al., 2008
Ascomycota	Sordariomycetes	Coniochaetaceae	*Coniochaeta (Lecythophora)*	*hoffmannii*	Pozo et al., 2011
Ascomycota	Sordariomycetes	Coniochaetaceae	*Coniochaeta*	*leucoplaca*	Pozo et al., 2012
Ascomycota	Taphrinomycetes	Taphrinaceae	*Taphrina*	*carpini*	Jacquemyn et al., 2013b
Basidiomycota	Incertae sedis	Incertae sedis	*Moniliella*	*megachiliensis*	Álvarez-Pérez and Herrera, 2013
Basidiomycota	Agaricostilbomycetes	Agaricostilbaceae	*Bensingtonia*	*miscanthi*	Mushtaq et al., 2007
Basidiomycota	Cystobasidiomycetes	Erythrobasidiaceae	*Erythrobasidium*	*hasegawianum*	Jacquemyn et al., 2013b
Basidiomycota	Microbotryomycetes	Incertae sedis	*Rhodoturula*	*aurantiaca*	Jacquemyn et al., 2013b
Basidiomycota	Microbotryomycetes	Incertae sedis	*Rhodotorula*	*colostri*	Pozo et al., 2011
Basidiomycota	Microbotryomycetes	Incertae sedis	*Rhodotorula*	*glutinis*	Golonka and Vilgalys, 2013
Basidiomycota	Microbotryomycetes	Incertae sedis	*Rhodotorula*	*graminis*	de Vega and Herrera, 2012
Basidiomycota	Microbotryomycetes	Incertae sedis	*Rhodotorula*	*fujisanensis*	de Vega and Herrera, 2012
Basidiomycota	Microbotryomycetes	Incertae sedis	*Rhodotorula*	*fragaria*	Mushtaq et al., 2007
Basidiomycota	Microbotryomycetes	Incertae sedis	*Rhodotorula*	*hinnulea*	Mushtaq et al., 2007
Basidiomycota	Microbotryomycetes	Incertae sedis	*Rhodotorula*	*mucilaginosa*	Pozo et al., 2011; 2012

Table 1. (Continued)

Phylum	Class	Family	Genus	Species	Reference
Basidiomycota	Microbotryomycetes	Incertae sedis	*Rhodoturula*	*nothofagi*	Álvarez-Pérez and Herrera, 2012
					de Vega and Herrera, 2012
Basidiomycota	Microbotryomycetes	Leucosporidiaceae	*Leucosporidiella*	*fragaria*	Jacquemyn et al., 2013b
Basidiomycota	Microbotryomycetes	Sporobolomycetaceae	*Sporobolomyces*	*phaffii*	Golonka and Vilgalys, 2013
Basidiomycota	Microbotryomycetes	Sporobolomycetaceae	*Sporobolomyces*	*roseus*	Álvarez-Pérez and Herrera, 2013
					de Vega and Herrera, 2012
					Jacquemyn et al., 2013a; 2013b
					Pozo et al., 2011
Basidiomycota	Microbotryomycetes	Sporobolomycetaceae	*Sporobolomyces*	*ruberrimus*	Golonka and Vilgalys, 2013
Basidiomycota	Tremellomycetes	Cystofilobasidiaceae	*Cystofilobasidium*	*bisporidii*	Mushtaq et al., 2008
Basidiomycota	Tremellomycetes	Cystofilobasidiaceae	*Cystofilobasidium*	*capitatum*	Brysch-Herzberg, 2004
Basidiomycota	Tremellomycetes	Cystofilobasidiaceae	*Mrakia*	*frigida*	Mushtaq et al., 2007; 2008
Basidiomycota	Tremellomycetes	Cystofilobasidiaceae	*Phaffia*	*rhodozyma*	Mushtaq et al., 2007; 2008
Basidiomycota	Tremellomycetes	Sirobasidiaceae	*Fibulobasidium*	*inconspicuum*	Mushtaq et al., 2007; 2008
Basidiomycota	Tremellomycetes	Tremellaceae	*Bullera*	*megalospora*	Mushtaq et al., 2007
Basidiomycota	Tremellomycetes	Tremellaceae	*Bullera*	*pseudoalba*	Mushtaq et al., 2008
Basidiomycota	Tremellomycetes	Tremellaceae	*Bullera*	*pyricola*	Mushtaq et al., 2007; 2008
Basidiomycota	Tremellomycetes	Tremellaceae	*Cryptococcus*	*aerius*	Pozo et al., 2011
Basidiomycota	Tremellomycetes	Tremellaceae	*Cryptococcus*	*aff. taibaiensis*	Canto and Herrera, 2012

Phylum	Class	Family	Genus	Species	Reference
Basidiomycota	Tremellomycetes	Tremellaceae	*Cryptococcus*	*aff. victoriae*	Jacquemyn et al., 2013a; 2013b
Basidiomycota	Tremellomycetes	Tremellaceae	*Cryptococcus*	*albidus*	Mushtaq et al., 2007; 2008
Basidiomycota	Tremellomycetes	Tremellaceae	*Cryptococcus*	*carnescens*	de Vega and Herrera, 2012
Basidiomycota	Tremellomycetes	Tremellaceae	*Cryptococcus*	*curvatus*	Mushtaq et al., 2007; 2008
Basidiomycota	Tremellomycetes	Tremellaceae	*Cryptococcus*	*diffluens*	Pozo et al., 2011
Basidiomycota	Tremellomycetes	Tremellaceae	*Cryptococcus*	*dimennae*	Golonka and Vilgalys, 2013
Basidiomycota	Tremellomycetes	Tremellaceae	*Cryptococcus*	*flavescens*	Golonka and Vilgalys, 2013
Basidiomycota	Tremellomycetes	Tremellaceae	*Cryptococcus*	*flavus*	Mushtaq et al., 2008
Basidiomycota	Tremellomycetes	Tremellaceae	*Cryptococcus*	*heveanensis*	Mushtaq et al., 2008
Basidiomycota	Tremellomycetes	Tremellaceae	*Cryptococcus*	*humicolus*	Mushtaq et al., 2008
Basidiomycota	Tremellomycetes	Tremellaceae	*Cryptococcus*	*hungaricus*	Mushtaq et al., 2008
Basidiomycota	Tremellomycetes	Tremellaceae	*Cryptococcus*	*laurentii*	Álvarez-Pérez and Herrera, 2013
					Canto and Herrera, 2012
					Golonka and Vilgalys, 2013
					Mushtaq et al., 2007; 2008
Basidiomycot	Tremellomycetes	Tremellaceae	*Cryptococcus*	*liquefaciens*	Canto and Herrera, 2012
Basidiomycot	Tremellomycetes	Tremellaceae	*Cryptococcus*	*luteolus*	Álvarez-Pérez and Herrera, 2013
Basidiomycota	Tremellomycetes	Tremellaceae	*Cryptococcus*	*macerans*	Álvarez-Pérez and Herrera, 2013
					de Vega and Herrera, 2012
					Golonka and Vilgalys, 2013
					Jacquemyn et al., 2013a

Table 1. (Continued)

Phylum	Class	Family	Genus	Species	Reference
					Mushtaq et al., 2008
Basidiomycota	Tremellomycetes	Tremellaceae	*Cryptococcus*	*magnus*	Golonka and Vilgalys, 2013
Basidiomycota	Tremellomycetes	Tremellaceae	*Cryptococcus*	*oeirensis*	Golonka and Vilgalys, 2013
					Jacquemyn et al., 2013a
Basidiomycota	Tremellomycetes	Tremellaceae	*Cryptococcus*	*stepposus*	Jacquemyn et al., 2013b
Basidiomycota	Tremellomycetes	Tremellaceae	*Cryptococcus*	*tephrensis*	Jacquemyn et al., 2013a
Basidiomycota	Tremellomycetes	Tremellaceae	*Cryptococcus*	*uzbekistanensis*	Pozo et al., 2011
Basidiomycota	Tremellomycetes	Tremellaceae	*Cryptococcus*	*victoriae*	Brysch-Herzberg, 2004
					Golonka and Vilgalys, 2013
					Jacquemyn et al., 2013a; 2013b
					Pozo et al., 2011
Basidiomycota	Tremellomycetes	Tremellaceae	*Holtermanniella*	*takashimae*	Jacquemyn et al., 2013a
Basidiomycota	Ustilaginomycetes	Ustilaginaceae	*Pseudozyma*	*antarctica*	Mushtaq et al., 2008
Basidiomycota	Ustilaginomycetes	Ustilaginaceae	*Pseudozyma*	*fusiformata*	Mushtaq et al., 2008
			Pseudozyma	*graminicola*	Golonka and Vilgalys, 2013
Basidiomycota	Ustomycetes	Sporidiobolaceae	*Rhodosporidium*	*toruloides*	Mushtaq et al., 2007; 2008
Basidiomycota	Ustomycetes	Sporidiobolaceae	*Sporidiobolus*	*ruineniae*	Mushtaq et al., 2007; 2008

Table 2. Overview of culturable bacteria found in floral nectar

Phylum	Class	Family	Genus	Reference
Actinobacteria	Actinobacteria	Actinosynnemataceae	Actinosynnemataceae sp.	Álvarez-Pérez and Herrera, 2013
Actinobacteria	Actinobacteria	Brevibacteriaceae	*Brevibacterium*	Jacquemyn et al., 2013a
Actinobacteria	Actinobacteria	Cellulomonadaceae	*Cellulomonas*	Álvarez-Pérez and Herrera, 2013
Actinobacteria	Actinobacteria	Dermabacteraceae	*Brachybacterium*	Jacquemyn et al., 2013a
Actinobacteria	Actinobacteria	Dermacoccaceae	*Flexivirga*	Jacquemyn et al., 2013a
Actinobacteria	Actinobacteria	Dermacoccaceae	*Luteipulveratus*	Jacquemyn et al., 2013a
Actinobacteria	Actinobacteria	Dermococcaceae	*Dermacoccus*	Jacquemyn et al., 2013b
Actinobacteria	Actinobacteria	Gordoniaceae	*Gordonia*	Jacquemyn et al., 2013a
Actinobacteria	Actinobacteria	Intrasporangiaceae	*Janibacter*	Jacquemyn et al., 2013a
Actinobacteria	Actinobacteria	Microbacteriaceae	*Curtobacterium*	Álvarez-Pérez and Herrera, 2013
				Fridman et al., 2012
				Jacquemyn et al., 2013b
Actinobacteria	Actinobacteria	Microbacteriaceae	*Frigoribacterium*	Jacquemyn et al., 2013b
Actinobacteria	Actinobacteria	Microbacteriaceae	*Leifsonia*	Álvarez-Pérez and Herrera, 2013
				Álvarez-Pérez et al., 2012
Actinobacteria	Actinobacteria	Microbacteriaceae	Microbacteriaceae sp.	Álvarez-Pérez and Herrera, 2013
Actinobacteria	Actinobacteria	Microbacteriaceae	*Microbacteriium*	Álvarez-Pérez et al., 2012
				Aizenberg-Gershtein et al., 2013
				Álvarez-Pérez and Herrera, 2013
				Jacquemyn et al., 2013a; 2013b
Actinobacteria	Actinobacteria	Microbacteriaceae	*Okibacterium*	Jacquemyn et al., 2013a

Table 2. (Continued)

Phylum	Class	Family	Genus	Reference
Actinobacteria	Actinobacteria	Microbacteriaceae	*Plantibacter*	Jacquemyn et al., 2013a
Actinobacteria	Actinobacteria	Micrococcaceae	*Arthrobacter*	Álvarez-Pérez and Herrera, 2013
				Fridman et al., 2012
				Jacquemyn et al.,2013a
Actinobacteria	Actinobacteria	Micrococcaceae	*Kocuria*	Álvarez-Pérez and Herrera, 2013
				Fridman et al., 2012
Actinobacteria	Actinobacteria	Micrococcaceae	*Micrococcus*	Álvarez-Pérez et al., 2012
				Jacquemyn et al., 2013a
Actinobacteria	Actinobacteria	Nakamurellaceae	*Saxeibacter*	Jacquemyn et al., 2013a
Actinobacteria	Actinobacteria	Nocardiaceae	*Rhodococcus*	Jacquemyn et al., 2013a
Actinobacteria	Actinobacteria	Nocardioidaceae	*Nocardioides*	Jacquemyn et al., 2013a
Actinobacteria	Actinobacteria	Propionibacteriaceae	*Ponticoccus*	Jacquemyn et al., 2013a
Actinobacteria	Actinobacteria	Streptomycetaceae	*Streptomyces*	Álvarez-Pérez and Herrera, 2013
				Jacquemyn et al., 2013a
Bacteroidetes	Sphingobacteriia	Chitinophagaceae	*Terrimonas*	Jacquemyn et al., 2013b
Proteobacteria	Alphaproteobacteria	Acetobacteraceae	Acetobacteraceae sp.	Álvarez-Pérez and Herrera, 2013
Proteobacteria	Alphaproteobacteria	Acetobacteraceae	*Asaia*	Álvarez-Pérez et al., 2012
				Fridman et al., 2012
Proteobacteria	Alphaproteobacteria	Acetobacteraceae	*Gluconobacter*	Vanette et al., 2013
Proteobacteria	Alphaproteobacteria	Bartonellaceae	*Bartonella*	Fridman et al., 2012
Proteobacteria	Alphaproteobacteria	Hyphomicrobiaceae	*Devosia*	Jacquemyn et al., 2013a
Proteobacteria	Alphaproteobacteria	Methylobacteriaceae	*Methylobacterium*	Álvarez-Pérez and Herrera, 2013
				Álvarez-Pérez et al., 2012
				Jacquemyn et al., 2013a; 2013b

Phylum	Class	Family	Genus	Reference
Proteobacteria	Alphaproteobacteria	Acetobacteraceae	*Gluconobacter*	Vanette et al., 2013
Proteobacteria	Alphaproteobacteria	Bartonellaceae	*Bartonella*	Fridman et al., 2012
Proteobacteria	Alphaproteobacteria	Hyphomicrobiaceae	*Devosia*	Jacquemyn et al., 2013a
Proteobacteria	Alphaproteobacteria	Methylobacteriaceae	*Methylobacterium*	Álvarez-Pérez and Herrera, 2013
				Álvarez-Pérez et al., 2012
				Jacquemyn et al., 2013a; 2013b
Proteobacteria	Alphaproteobacteria	Rhizobiaceae	Rhizobiaceae sp.	Álvarez-Pérez and Herrera, 2013
Proteobacteria	Alphaproteobacteria	Rhizobiaceae	*Rhizobium*	Álvarez-Pérez and Herrera, 2013
Proteobacteria	Alphaproteobacteria	Sphingomonadaceae	*Sphingomonas*	Álvarez-Pérez and Herrera, 2013
				Álvarez-Pérez et al., 2012
				Jacquemyn et al., 2013b
Firmicutes	Bacilli	Bacillaceae	*Bacillus*	Aizenberg-Gershtein et al., 2013
				Álvarez-Pérez and Herrera, 2013
				Álvarez-Pérez et al., 2012
				Fridman et al., 2012
				Jacquemyn et al., 2013a; 2013b
Firmicutes	Bacilli	Bacillaceae	*Terribacillus*	Aizenberg-Gershtein et al., 2013
Firmicutes	Bacilli	Enterococcaceae	*Enterococcus*	Álvarez-Pérez and Herrera, 2013
Firmicutes	Bacilli	Leuconostocaceae	*Leuconostoc*	Jacquemyn et al., 2013b
				Álvarez-Pérez and Herrera, 2013
Firmicutes	Bacilli	Paenibacillaceae	*Paenibacillus*	Álvarez-Pérez et al., 2012
				Fridman et al., 2012
				Jacquemyn et al., 2013b

Table 2. (Continued)

Phylum	Class	Family	Genus	Reference
Firmicutes	Bacilli	Staphylococcaceae	*Staphylococcus*	Aizenberg-Gershtein et al., 2013
				Álvarez-Pérez and Herrera, 2013
				Fridman et al., 2012
				Jacquemyn et al., 2013a; 2013b
Firmicutes	Bacilli	Streptococcaceae	*Lactococcus*	Álvarez-Pérez and Herrera, 2013
Proteobacteria	Betaproteobacteria	Alcaligenaceae	Alcaligenaceae sp.	Álvarez-Pérez and Herrera, 2013
				Álvarez-Pérez et al., 2012
Proteobacteria	Betaproteobacteria	Burkholderiaceae	*Burkholderia*	Álvarez-Pérez et al., 2012
Proteobacteriaceae	Betaproteobacteria	Comamonadaceae	*Variovorax*	Álvarez-Pérez and Herrera, 2013
Bacteroidetes	Flavobacteriia	Flavobacteriaceae	*Cryseobacterium*	Fridman et al., 2012
Bacteroidetes	Flavobacteriia	Flavobacteriaceae	Flavobacteriaceae sp.	Álvarez-Pérez and Herrera, 2013
Proteobacteria	Gammaproteobacteria	Enterobacteriaceae	*Enterobacter*	Jacquemyn et al., 2013b
Proteobacteria	Gammaproteobacteria	Enterobacteriaceae	Enterobacteriaceae sp.	Álvarez-Pérez and Herrera, 2013
				Álvarez-Pérez et al., 2012
Proteobacteria	Gammaproteobacteria	Enterobacteriaceae	*Erwinia*	Fridman et al., 2012
				Jacquemyn et al., 2013a; 2013b
Proteobacteria	Gammaproteobacteria	Enterobacteriaceae	*Pantoea*	Aizenberg-Gershtein et al., 2013
				Álvarez-Pérez et al., 2012
				Fridman et al., 2012
Proteobacteria	Gammaproteobacteria	Enterobacteriaceae	*Pectobacterium*	Jacquemyn et al., 2013b

Phylum	Class	Family	Genus	Reference
Proteobacteria	Gammaproteobacteria	Enterobacteriaceae	*Plesiomonas*	Jacquemyn et al., 2013b
Proteobacteria	Gammaproteobacteria	Enterobacteriaceae	*Rhanella*	Jacquemyn et al., 2013b
Proteobacteria	Gammaproteobacteria	Enterobacteriaceae	*Rosenbergiella*	Jacquemyn et al., 2013b
				Aizenberg-Gershtein et al., 2013
				Fridman et al., 2012
				Halpern et al., 2013
				Lenaerts et al., 2014
Proteobacteria	Gammaproteobacteria	Enterobacteriaceae	*Serratia*	Jacquemyn et al., 2013b
Proteobacteria	Gammaproteobacteria	Enterobacteriaceae	*Tatumella*	Jacquemyn et al., 2013b
Proteobacteria	Gammaproteobacteria	Moraxellaceae	*Acinetobacter*	Aizenberg-Gershtein et al., 2013
				Álvarez-Pérez and Herrera, 2013
				Álvarez-Pérez et al., 2013b
				Fridman et al., 2012
				Jacquemyn et al., 2013b
Proteobacteria	Gammaproteobacteria	Moraxellaceae	*Moraxella*	Jacquemyn et al., 2013a
Proteobacteria	Gammaproteobacteria	Pseudomonadaceae	*Pseudomonas*	Álvarez-Pérez et al., 2012
				Jacquemyn et al., 2013a; 2013b
				Fridman et al., 2012
				Álvarez-Pérez and Herrera, 2013
				Álvarez-Pérez et al., 2013a
Proteobacteria	Gammaproteobacteria	Xanthomonadaceae	*Rhodanobacter*	Jacquemyn et al., 2013a
Proteobacteria	Gammaproteobacteria	Xanthomonadaceae	*Stenotrophomonas*	Álvarez-Pérez et al., 2012
Proteobacteria	Gammaproteobacteria	Xanthomonadaceae	Xanthomonadaceae sp.	Álvarez-Pérez et al., 2012

Floral nectar is assumed to be initially sterile (Brysch-Herzberg, 2004; Canto et al., 2008). However, some microorganisms, mainly yeasts and bacteria, may rapidly colonize the nectar by inoculation from the air or by making contact with the flower corolla or the proboscis of pollinators such as bumblebees, some birds, and ants (Brysch-Herzberg, 2004; de Vega et al., 2009; Herrera et al., 2009; Pozo et al., 2011; 2012; de Vega & Herrera, 2013; Aizenberg-Gershtein et al., 2013). Microbes have been isolated from flowers and more specifically, from nectar samples, all over the world (Table 1 and 2).

Although yeast contamination rates of nectar samples can be high (de Vega et al., 2009, Herrera et al., 2009; Belisle et al., 2011), nectar-associated yeast communities are generally species-poor (on average 1.2 culturable yeast species/sample) and mostly dominated by the insect- or bird-vectored yeast genus *Metschnikowia* (Saccharomycetales) (Canto et al., 2008; de Vega et al., 2009; Belisle et al., 2012, Pozo et al., 2012). Other yeasts recorded include, amongst many others, species of the genera *Rhodotorula*, *Cryptococcus*, *Sporobolomyces*, and *Candida* although they generally occur in much lower frequencies (Table 1). Most species within these genera commonly occur in flowers (Sandhu & Waraich, 1985), and may become inoculated into the nectar by, for example, contact with the corolla or the visiting insects, despite not being truly specialized in nectar (Pozo et al., 2012).

In addition to yeasts, bacteria have been shown to be common inhabitants of floral nectar. The overall frequency of occurrence of bacteria in plant species ranged from 20% to 80% in Mediterranean and South African plant species, respectively (Álvarez-Pérez et al., 2012, Álvarez-Pérez & Herrera, 2013). By analogy with yeasts, bacterial species richness is also low (on average 1.4 bacterial species per nectar sample) (Álvarez-Pérez and Herrera, 2013), with most bacteria belonging to the Actinobacteria, Firmicutes, and Proteobacteria (Table 2) (Álvarez-Pérez et al., 2012; Fridman et al., 2012; Jacquemyn et al., 2013a; Jacquemyn et al., 2013b). Strikingly, these bacteria often co-occur with yeasts such as *Metschnikowia* (Álvarez-Pérez & Herrera, 2013), perhaps displaying complementary nutrient utilisation patterns (Álvarez-Pérez et al., 2013b). Recently, a number of novel bacterial species isolated from floral nectar, all belonging to the Gammaproteobacteria, have been described, including two new species of the family Moraxellaceae, *Acinetobacer nectaris* and *A. boissieri* (Alvarez-Perez et al., 2013b), and four of the family Enterobacteriaceae, including *Rosenbergiella nectarea* (Halpern et al., 2013), *R. australoborealis R. collisarenosi* and *R. epipactidis* (Lenaerts et al., 2014).

Factors Structuring Microbial Communities in Floral Nectar

Most research so far has indicated that nectar-dwelling microorganisms are generally organized as phylogenetically clustered species-poor communities in comparison with the potential pool of insect-vectored microbial colonizers (Herrera et al., 2010; Peay et al., 2012), suggesting that there may be important filtering mechanisms that limit species diversity in nectar. According to the 'nectar-filtering hypothesis', the high sugar content (and associated low water activity) and the presence of plant secondary compounds that typically characterize floral nectar may have a strong impact on the survival and proliferation of immigrating microbes (Adler, 2000; Herrera et al., 2010). As a result, different microbial communities consisting of the best adapted strains will establish in different plant species (Herrera et al., 2010). However, tolerance to plant secondary compounds such as alkaloids or glycosides was not found to be a major determinant structuring microbial communities in nectar as many plant compounds did not show any measurable inhibitory effect on the survival of several potential nectar colonizers (Pozo et al., 2012, but see Manson et al., 2007). A stronger filtering effect may be expected from the high sugar concentrations typically encountered in nectar. Nectar typically represents an intermediate or high water-activity habitat, having a_w levels ranging from 0.90 to >0.99 a_w (Ferro Fontán & Chirife, 1981; Nicolson, 1994; 2002), and may therefore constrain microbial growth of most microbial cells that cannot cope with (relatively) high osmotic pressure (Lievens et al., 2014). As a result, basidiomycetous yeasts frequently present in the phylloplane, fresh water, or air (Fonseca & Inácio, 2004) are considered secondary nectar inhabitants, and are probably not the best adapted species to persist in the harsh environment of nectar (Pozo et al., 2012). However, low water-activity specialized genera such as *Starmerella*, *Debaryomyces*, and *Zygosaccharomyces* are also not frequently isolated from floral nectar, while they have been frequently found in other related habitats such as bee nest provisions (Pozo et al., 2012, Rosa et al., 2003, Pozo, Lachance & Herrera, unp. results). Species of *Metschnikowia*, on the other hand, which can deal with sugar concentrations up to 50% (corresponding to a_w = 0.87, Sereno et al., 2001), are the most commonly found yeasts in nectar (87% of isolates coming from nectar individual samples belonging to 25 plant species, Pozo et al., 2011).

Metschnikowia strains are typically adapted to relatively high C/N contents and differ from species found in the phylloplane and air in their nitrogen and carbon assimilation patterns (Pozo et al., 2012, Peay et al., 2013). In addition, fermentation may occur in nectar (Wiens et al., 2008), implying that oxygen availability and accumulation of ethanol might be considered additional growth-limiting environmental factors. *Metschnikowia* shows fast sucrose and glucose assimilation, and high ethanol resistance (Pozo et al., 2012). However, it is reasonable to assume that proliferation of *Metschnikowia* is to a certain extent also affected by temperature, as temperature is generally known to exert a major influence on biological activity and microbial growth (Lachance, 2006). Given that temperature is a highly variable feature at different temporal scales and that nectar acts as discrete and open habitats that are constrained by floral lifespan and pollinator consumption of nectar resources, it can be expected that nectar-inhabiting species should show rapid biomass development at a wide temperature range. Experiments have indeed shown that several *Metschnikowia* strains (from the species *M. kunwiensis*, *M. reukaufii* and *M. gruessii*) can grow at fast rates at temperature ranges between 8 and 30ºC (Pozo et al., 2012).

Due their ephemeral nature, communities of nectar microorganisms will inevitably function as metacommunities (i.e., a set of local communities that exchange species) with high patch turnover rates and intense colonization-extinction dynamics. As nectar chemistry can vary strongly within and between different plant species in which the microbes become inoculated (Nicholson & Thornbrug, 2007; Herrera et al., 2006), proliferation also depends on the ability of microorganisms to cope with a broad range of nectar environments and thus requires the ability to rapidly adapt to different nectar conditions. In highly specific and contrasting habitats natural selection may favour the settlement of extreme or unusual genotypes, a process known as "diversifying selection". This kind of selection was postulated as the key mechanism maintaining high genotypic diversity in *Metschnikowia*, as those isolates explore different niches during their life cycle (Herrera et al., 2011; Herrera et al., 2014). Furthermore, epigenetic changes in the microbes also contribute to population niche width by enhancing phenotypic plasticity. In the specific case of the nectar specialist *M. reukauffi*, genome-wide DNA methylation patterns enable this species to exploit a broad sugar concentration and sugar type range (Herrera et al., 2012).

However, dispersion and historic events and the order of species arrival may additionally explain the relative abundance and dynamics of yeasts and bacteria in nectar (Peay et al., 2012; Lievens et al., 2014; Tucker & Fukami,

2014). For instance, as yeasts and bacteria have the potential to change sugar concentrations (see further), this may enable later colonization by less specialized species and, for example, explain why some basidiomycetous taxa can be encountered in floral nectar. Additionally, nectar represents a habitat with potential for strong priority effects, where the outcome of species interactions depends on the order of their arrival, as has recently been demonstrated for a number of nectar yeasts and bacteria (Peay et al., 2012). Furthermore, it has been shown that environmental variability, mimicked by spatial and temporal temperature shifts, may have an influence on the strength of priority effects. More specifically, temperature variability was found to prevent extinction of late-arriving species that would have been excluded owing to priority effects if temperature had been constant (Tucker & Fukami, 2014).

IMPACT OF MICROORGANISMS ON FLORAL NECTAR

The presence of nectar-dwelling microorganisms can have a huge impact on nectar chemistry, especially on nectar pH and sugar composition and concentration (Figure 1; Canto & Herrera, 2012, Vannette et al., 2013), the concentration of amino acids (Vannette et al., 2013), nectar temperature and odour (Herrera & Pozo, 2010; Golonka et al., 2014). For example, in the hummingbird pollinated shrub *Mimulus aurantiacus*, both experimentally inoculated *Gluconobacter* and *Metschnikowia* (10^4 cells/mm^3 cultures, added to artificial nectar) strains reduced H_2O_2 concentrations by about 80% (Vannette et al., 2013). *Gluconobacter* decreased pH by 5 units (a 10^5 increase in H^+ concentration) and total sugar and sucrose concentration by 27 and 35%, respectively, whereas *M. reukaufii* decreased pH by 2 units and total sugar and sucrose concentration by 16 and 17%, respectively. *Gluconobacter* reduced glucose concentration by 64% and increased fructose concentration by 42%, whereas *M. reukaufii* had little effect on glucose and fructose concentration (Vannette et al., 2013). However, other studies consistently relate the presence of ascomycetous yeasts in nectar with significant changes in the proportion of sucrose, glucose and fructose, by reducing sucrose concentration and leaving unbalanced proportions of their monomers, with a high preponderance of fructose in natural nectar samples (Figure 1; Herrera et al., 2008; de Vega et al., 2009; Canto & Herrera, 2012).

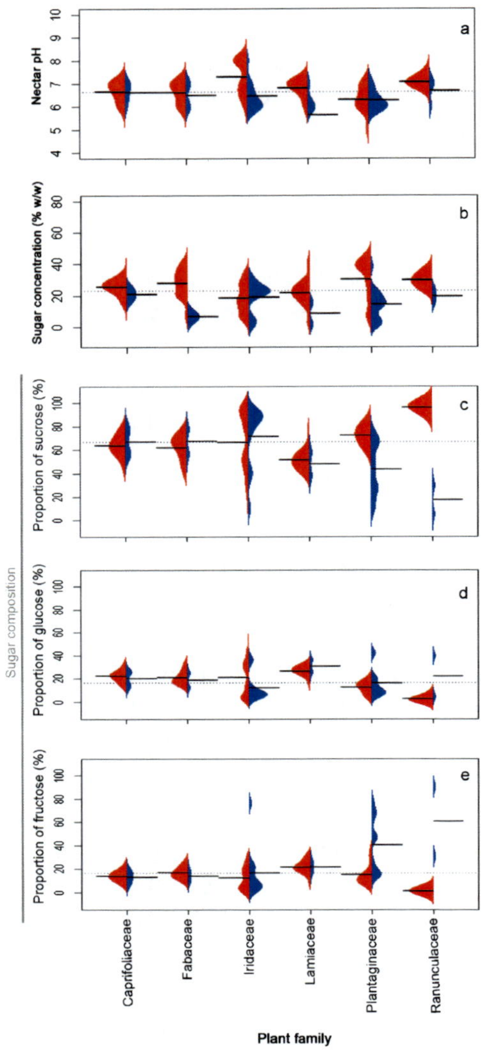

Figure 1. Beanplots showing differences between virgin (sterile) floral nectar (left, red) and nectar isolated from plants that could be accessed by insects (potentially contaminated with microorganisms, right, blue) for different nectar traits: (a) pH; (b) total sugar concentration (sucrose, glucose and fructose) (w/w); and relative proportions (%) of (c) sucrose, (d) glucose and (e) fructose compared to the total of these three main nectar sugars. Virgin nectar was obtained from plants protected from insect visitation. Data were obtained from 16 plant species belonging to 6 plant families (x axis). Beanplots are an alternative to boxplots for visual comparison of univariate data between groups. A bean consists of a one-dimensional scatter plot, its distribution as a density shape and an average line for the distribution. The overall average for the plot is drawn as a line. Figure taken from Lievens et al. (2014).

Canto and Herrera (2012) investigated the relationship between sugar concentration and yeast density in a wide number of plant species and found that nectar fructose, glucose and sucrose significantly declined with increasing yeast density in certain plant species, but not in others, indicating that changes in sugar concentration associated with yeast density are to some extent species-specific. The combined effect of plant species and yeast density on nectar chemistry might be partially accounted by the fact that different plant species may harbour different yeast communities, as several yeast species were isolated in their study.

Next to changes in sugar concentration and composition, nectar microbes may also change the amino acid content in nectar (Peay et al., 2012). The nectar-inhabiting yeasts *Candida floricola* and *Metschnikowia koreensis, M. kunwiensis* and *M. reukaufii* have been shown to decrease significantly the concentration of the non-essential amino acids proline, glutamic acid and aspartic acid in the nectar of *M. aurantiacus* (Peay et al., 2012). Amino acids may also diffuse in nectar from dehiscent pollen grains (Gilbert, 1972). For instance, pollen grain walls, which are usually strong (Heslop-Harrison & Heslop-Harrison, 1988), may be disrupted in nectar after which their content is released (Pozo, pers. observation). As a result, nectar yeasts as *M. reukafiii* tend to aggregate in the vicinity of pollen grains fallen into the nectar (Figure 2b,d), leading to a positive correlation between the density of pollen grains and yeast density in the nectar (Figure 2a). Although nectar yeast cell density and pollen density vary among plant species, yeast densities in nectar were highest in plant families with the highest pollen densities, indicating that the presence and abundance of pollen in nectar positively affects yeast densities (Figure 2c). These findings suggest that nitrogen may be the main factor limiting growth of nectar-inhabiting yeasts, as amino acid concentrations in newly secreted nectar are generally quite low (Baker & Baker, 1973; 1986).

Besides reducing sugar and amino acid concentrations and changing their composition, the metabolic activity of nectar-inhabiting microorganisms can also affect other floral attractive traits, including temperature (the fermentative and fermentative–oxidative metabolic activity of yeasts produces significant amounts of heat (Herrera & Pozo, 2010)), and scent (Raguso 2004; Goodrich et al., 2006). Herrera and Pozo (2010), for example, showed that experimental addition of yeasts to the nectar of the winter-blooming *Helleborus foetidus* significantly increased the temperature excess ($\Delta T_{nect} = T_{nect} - T_{air}$) of nectaries. In non-experimental flowers, ΔT_{nect} significantly increased with increasing yeast density, with a temperature excess of 6 °C at yeast densities > 10^5 cells/mm^3.

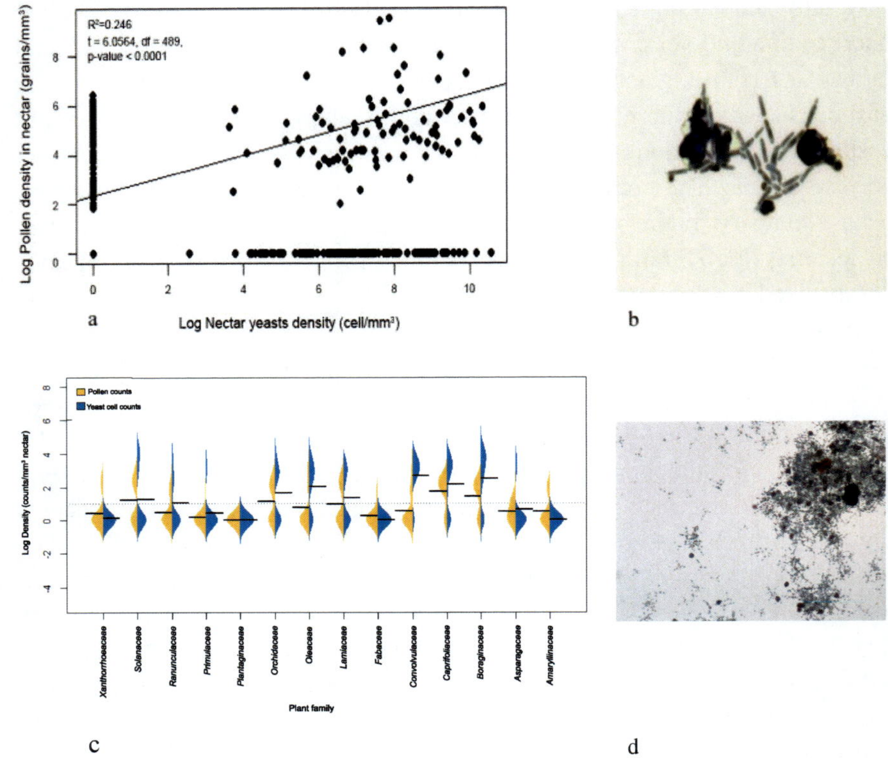

Figure 2. (a) Correlation between yeast density and pollen density in 491 microscopically examined nectar samples, belonging to 21 plant species and 14 families. (b) *Metschnikowia reukaufii* and pollen grains co-occurring in in nectar of *Helleborus foetidus*, stained with Lactophenol cotton blue, 40x amplification. (c) Beanplots showing differences between density of pollen grains (left, yellow) and yeast density (right, blue) for 491 nectar samples, belonging to 21 plant species and 14 plant families (x axis). (d) Nectar sample of *Teucrium pseudochamaepytis*, stained with Lactophenol cotton blue, showing yeast cells and pollen grains at10X amplification.

Warmer nectar could offer energetic advantages for insect thermos-regulation, as well as being easier to drink owing to its lower viscosity (Nicolson et al., 2013). Some nectars have been shown to be scented (Raguso, 2008). While some VOCs might be diffused from the floral surroundings, recent evidence also points to nectar-inhabiting microorganisms as responsible of nectar scents (Golonka et al., 2014). By comparing the scent profiles of unvisited (sterile) nectar, unvisited flowers, and visited (microbial contaminated) nectar the authors confirmed that *Silene caroliniana* only produces a few scented compounds, while many more were produced when microbes were present in the nectar. In particular, nectar-inhabiting *Metschnikowia* species produced aliphatic alcohols, including ethanol, 2-methyl-1-propanol, 3-methyl-1-butanol, and 2-methyl-1-butanol, contributing to *S. caroliniana*´s floral scent.

Impact of Microorganisms on Pollinator Behaviour and Plant Fitness

Because nectar microorganisms can have a major impact on nectar quality as resource for pollinators (Canto et al., 2007; Canto et al., 2008; de Vega et al., 2009; Herrera et al., 2008; Herrera & Pozo, 2010; Peay et al., 2011), it is reasonable to assume that they also affect pollinator behavior and ultimately plant fitness. Ehlers and Olesen (1997), for example, showed that the presence of some microbes in the nectar of *Epipactis helleborine* was associated with high ethanol concentrations in nectar, suggesting that the metabolic action of microbes caused the singular behavior of wasps foraging on those plants. On the other hand, Kevan et al. (1988) showed that foraging honeybees did not discriminate between yeast-contaminated and yeast-free flowers. However, neither yeast density nor nectar characteristics were evaluated in that study.

The ecological importance of nectar-associated microorganisms for plant-pollinator mutualisms was only recently explicitly acknowledged in three independently conducted studies (Herrera et al., 2013, Vannette et al., 2013; Schaeffer & Irwin, 2014). The presence of the nectar specialist *M. reukaufii* led to a significant increase in bumblebee flower visits of *Helleborus foetidus*, either in controlled assays or field experiments (Herrera et al., 2013), whereas the presence of *Gluconobacter* decreased nectar consumption and flower visits by hummingbirds in *Mimulus aurantiacus* (Vannette et al., 2013). High yeast densities in the nectar of the low larkspur *Delphinium nuttallianum* increased the amount of nectar removed by both bumblebee and hummingbird pollinators (Schaeffer & Irwin, 2014). The exact reasons why bumblebees and hummingbirds probe more flowers with high yeast densities are not entirely clear. It might be due to sensory cues such as smell, tactile cues, or physico-chemical features such as viscosity, and sugar or amino acid composition and concentration (Pozo, 2012; Herrera et al., 2013). Further, Good et al. (2014) showed that honey-bees avoided flowers with nectar colonized by the bacterial species *Asaia astilbis*, *Erwinia tasmaniensis*, and *Lactobacillus kunkeei*, most likely by altering nectar chemistry, whereas the yeast *M. reukaufii* did not affect the feeding preference of the insects.

Microbial communities inhabiting floral nectar can affect pollinator services and pollination efficiency not only by reducing rewards, but also in several other ways. First, yeast distribution in nectar samples may be extremely patchy, with variation in yeast density mostly occurring at the within-plant level (Pozo et al., 2014). As a result, flowers from the same plant often contain yeasts at different densities in their nectar, giving rise to a different composition of the nectar rewards at the within-plant level (Herrera et al., 2008; Pozo et al., 2014), which in turn may negatively affect risk-averse pollinators (Biernaskie et al., 2002). Second, fermenting yeasts produce ethanol (Lin & Tanaka, 2006; Hahn-Hägerdal et al., 2007), and the presence of alcohol in nectar could intoxicate pollinators and therefore alter their foraging behaviour (Ehlers & Olesen, 1997; Wiens et al., 2008). Third, yeasts are able to produce a wide range of scents (e.g., Majdak et al., 2002; Swiegers et al., 2005), which may also impact on pollinator behavior. In *Silene caroliniana*, for example, nectar-living *M. reukaufii* and *M. koreensis* contributed ethanol and other aliphatic alcohols to the overall flower scent (Golonka et al., 2014). Volatiles emissions by certain yeast species have been shown to attract beetles (Graham et al., 2011). Butterflies and moths were also shown to prefer artificial flowers containing scented nectar over those that contain pure sugar solutions (Weis, 2001). However, to what extent nectarivorous yeast can affect floral scent and filter potential pollinators through volatile release remains to be determined.

Few studies have investigated the impact of nectar microbiota on plant reproductive success. Eisikowitch et al. (1990) showed that the presence of *M. reukaufii* in the nectar of *Asclepias syriaca* decreased pollen germination. Sugar concentrations in the nectar of *Asclepias* must be in a certain range to assure pollen germination, decreases in sugar concentration due to microbial activity might be responsible for the decreased pollen germination. Somewhat similar results were provided by Herrera et al. (2013), who showed that experimental inoculation of the nectar of *H. foetidus* with the nectar specialist *M. reukaufii* interfered with pollen germination. More in particular, presence of *M. reukaufii* led to reductions in the number of pollen tubes in the style, fruit set, seed set, and individual seed mass, providing direct evidence that nectar yeasts can modify pollinator foraging patterns and thus pollination success. Addition of *M. reukaufii* to the nectar of *Delphinium nuttallianum* increased the amount of nectar removed by bumblebee pollinators, and it also enhanced male fitness component by increasing pollen donation. However, the addition of yeasts did not affect the female component of plant fitness, either measured by fruit set, number of seeds per fruit or total seeds per plant (Schaeffer & Irwin, 2014). Vannette et al. (2013), on the other hand, showed that nectar bacteria, rather than yeasts, reduced pollination success and seed set in *Mimulus aurantiacus*. Inoculation with *Gluconobacter* decreased the number of seeds produced by an average of 18 %, compared with the control treatment, whereas *M. reukaufii* inoculation did not significantly affect stigma closure or seed set compared with the control treatment. It seems that the effect of nectar-inhabiting microbes on plant fitness may depend on plant specific attributes such as flower morphology, plant mating system, the component of reproduction measured, and the pollen limitation experienced (Herrera et al., 2013), apart from its direct effect on pollinators (Schaefer & Irwin, 2014).

FUTURE PERSPECTIVES

Although it has become clear that nectar-inhabiting microorganisms can be expected to play an important role in the interaction between plants and pollinators, further research is needed to increase our understanding of the ecological role of nectar microbiota. First, since nectar-inhabiting microorganisms have to cope with constantly varying habitats during their life cycle, they can be expected to have broad ecological niches with a high degree of phenotypic plasticity (Roughgarden 1972; Sultan & Spencer 2002; Baythavong, 2011). At present, little is known about the phenotypic profile of nectar-inhabiting yeasts and bacteria. Preliminary results have shown that the nectar-living *Metschnikowia* species (*M. gruessii* and *M. reukaufii*) can explore a wide phenotypic landscape (Pozo et al., unpublished manuscript). Moreover, nectar-living yeast species have been shown to possess a high genotypic diversity, despite of reproducing mostly clonally in nature (Herrera et al., 2011, Herrera et al., 2014). Future research aiming to elucidate the effects of the broad phenotypic and genotypic diversity on the ecology of nectar microbes should carefully consider the origin and the number of strains per individual species in the experimental design.

Second, it is feasible to think that plants have evolved mechanisms to prevent microbial colonization in floral nectar to provide pollinators with pristine nectar (Adler, 2000; Carter & Thornburg, 2004). In this regard, several classes of antimicrobial proteins have been identified putatively protecting nectar from microbial invasion (Carter et al., 2007; Park & Thornburg, 2009;

Heil, 2011). For instance, several proteins are involved in the nectar redox cycle in *Nicotiana*, and the presence of lipases, RNases, GABA, and ABBA proteins have been documented in nectar as well. However, the antimicrobial effectiveness of many of them remains largely untested for nectar-inhabiting microorganisms (Nepi, 2014). Moreover, potentially toxic secondary compounds have been described in floral nectar from many plant species from different families, including alkaloids, phenolics, saponins, non-protein amino acids, and ammonia (high pH) (Adler, 2000; Manson et al., 2007), lactones (Mares, 1987) or glycosides (Pozo et al., 2012). Although toxic compounds may just be transported to the nectar via phloem, toxic nectar has also been given some ecological significance, by deterring nectar robbers. However, the antimicrobial efficacy of those compounds has, with a few exceptions (Golonka, 2002), not been fully proven experimentally for most plant species (Manson et al., 2007; Pozo et al., 2012). Better insights into the presence of secondary compounds and its concentrations in nectar of different plants are needed in order to perform more realistic assays on nectar toxicity, both aiming to mimic nectar complex conditions and to test either single compounds or mixtures. Physiological assays dealing with the solubility and stability of these compounds in nectar conditions might also explain why microbes could not be deterred by potentially toxic plant secondary compounds.

Third, so far, most studies on nectar microbial ecology aimed at better understanding the impact of nectar microbial communities on pollination and plant reproductive success, while the importance of nectarivorous microorganisms on other important flower-visiting insects such as nectar-consuming natural enemies of insect pests remains unexplored. However, by analogy with nectarivorous microorganisms that intervene in plant-pollinator mutualisms, it can be expected that nectar-inhabiting microorganisms also play a major role in the interactions between plants, natural enemies and pest insects. Given the increasing use of insect predators and parasitoids in pest control programs (Bale et al., 2008), often in combination with nectar-producing plants to provide the insects with the necessary sugars to cover their energetic needs (Zender et al., 2007; Johnson et al., 2008), a more global understanding of the ecological role of nectar-dwelling microorganisms may lead to enhanced biocontrol in agricultural cropping systems. In that sense, it would be useful to know if plant chemical defenses are induced by microbial inhabiting microorganisms. If so, nectar microbes may have cascading or reciprocal effects on the other organisms that interact with plants, including nectar robbers, insect pest or plant herbivores.

Fourth, pollinator fitness might also be enhanced or decreased by intake of the nectar-inhabiting microbes. It is already known that some nectar bacterial and yeast species inhabit the proboscis and guts of bees (Good et al., 2014; Pozo et al., unpublished results). Bees exclusively rely on plant derived food for nutrition. However, this food must be processed by enzymes and microbial flora in the guts of bees. Some of these microbes may help to process some complex macromolecules, such as polypeptides and polysaccharides, as is the case with proteobacteria inhabiting the honeybee gut (Lee et al., 2014). In addition, some main nectar yeast species leave fructose behind in nectar, a sugar that has been shown to increase sugar solution intake in artificial colonies of bumblebees (Pozo, 2012). Moreover, yeast inhabiting floral nectar may also provide, or represent by themselves some sort of reward to pollinators. For example, nectar yeast metabolic action may reward pollinators by warming the flowers of winter-flowering plants (Rands & Whitney, 2008; Herrera & Pozo, 2010), and yeast cells may provide a protein or amino acid source to bumblebees. Better insights into the role of insect-associated microbes on the insect life cycle are needed to better understand the nature of the tritrophic interactions between plants, microbials and insects.

Finally, it has already been shown that the nectar yeast *M. reukaufii* produces significant quantities of volatiles, which give a particular smell to nectar (Golonka et al., 2014). This suggests that microbial activity may affect "before-probing" nectar attractiveness to pollinators (González-Teuber and Heil, 2009). Current evidence about the differential behavior of artificial bumblebee colonies when specific yeasts species are present in nectar give rise to more complex experiments, such as tests under field conditions, or tests using different microbes, such as different yeast and bacterial taxa, and their mixtures. Therefore, future research is also needed to discern the detailed mechanisms underlying pollinator attraction to nectar infested with some microbes.

ACKNOWLEDGMENTS

MIP acknowledges financial support from the European Union program FP7 PEOPLE-2012-IEF (grant 327635), and HJ acknowledges funding from the European Research Council (ERC starting grant 260601 – MYCASOR). Thank John Wiley and Sons, Ltd., for granting permission to include Figure 1, taken from Lievens et al. (2014). Microbiology of sugar-rich environments:

diversity, ecology, and system constraints. Environmental Microbiology, Wiley and Sons, Ltd. in the printed and electronic version of this book.

REFERENCES

Adler, L. S. (2000). The ecological significance of toxic nectar. *Oikos, 91*, 409–420.

Aizenberg-Gershtein, Y., Izhaki, I., & Halpern, M. (2013). Do honeybees shape the bacterial community composition in floral nectar? *PLoS One, 8*, e67556.

Álvarez-Pérez, S., Herrera, C. M., & de Vega, C. (2012). Zooming-in on floral nectar: a first exploration of nectar-associated bacteria in wild plant communities. *FEMS Microbiology Ecology, 80*, 591-602.

Álvarez-Pérez, S., & Herrera, C. M. (2013). Composition, richness and non-random assembly of culturable bacterial microfungal communities in floral nectar of Mediterranean plants. *FEMS Microbiology Ecology, 83*, 685-699.

Álvarez-Pérez, S., de Vega, C., & Herrera, C. M. (2013a). Multilocus sequence analysis of nectar pseudomonads reveals high genetic diversity and contrasting recombination patterns. *PLoS One, 8*, e75797.

Álvarez-Pérez, S., Lievens, B., Jacquemyn, H., & Herrera, C. M. (2013b). *Acinetobacter nectaris* sp. nov. and *Acinetobacter boissieri* sp. nov., two novel bacterial species isolated from floral nectar of wild Mediterranean insect-pollinated plants. *International Journal of Systematic and Evolutionary Microbiology, 63*, 1532-1539.

Baker, H. G., & Baker, I. (1973). Aminoacids in nectar and its evolutionary significance. *Nature, 241*, 543-545.

Baker, H. G. & Baker, I. (1977). Intraspecific constancy of floral nectar amino acid complements. *Botanical Gazette, 138,* 183-191.

Baker, H. G. & Baker, I. (1983). Floral nectar sugar constituents in relation to pollinator type. In: C. Jones, & R. Little, *Handbook of Experimental Pollination Biology*. New York, Van Nostrand Reinhold.

Baker, H. G. & Baker, I. (1986). The occurrence and significance of amino acids in floral nectar. *Plant Systematics and Evolution, 151*, 175-186.

Bale, J. S., Van Lenteren, J. C., & Bigler, F. (2008). Biological control and sustainable food production. *Philosophical Transactions of the Royal Society, 363*, 761-776.

Baythavong, B. S. (2011). Linking the spatial scale of environmental variation and the evolution of phenotypic plasticity: selection favors adaptive plasticity in fine-grained environments. *The American Naturalist, 178*, 75-87.

Belisle, M., Peay, K., & Fukami, T. (2012). Flowers as islands: spatial distribution of nectar-inhabiting microfungi among plants of *Mimulus aurantiacus*, a hummingbird-pollinated shrub. *Microbial Ecology, 3*, 711-718.

Biernaskie, J. M., Cartar, R. V., & Hurly, T. A. (2002). Risk-averse inflorescence departure in hummingbirds and bumble bees: could plants benefit from variable nectar volumes? *Oikos, 98*, 98-104.

Boutroux, L. (1884). Conservation des ferments alcooliques dans la nature. *Annales des Sciences Naturelles; Botanique, 17,* 145-209.

Brysch-Herzberg, M. (2004). Ecology of yeasts in plant-bumblebee mutualism in Central Europe. *FEMS Microbiology Ecology, 50*, 87-100.

Canto, A., Pérez, R., Medrano, M., Castellanos, M. C., & Herrera, C. M. (2007). Intraplant variation in nectar sugar composition in two *Aquilegia* species (Ranunculaceae): contrasting patterns under field and greenhouse conditions. *Annals of Botany, 99*, 653-660.

Canto, A., Herrera, C. M., Medrano, M., Pérez, R., & García, I. M. (2008). Pollinator foraging modifies nectar sugar composition in *Helleborus foetidus* (Ranunculaceae): An experimental test. *American Journal of Botany, 95*, 315-320.

Canto, A., Herrera, C. M., García, I. M., Pérez, R., & Vaz, M. (2011). Intraplant variation in nectar traits in *Helleborus foetidus* (Ranunculaceae) as related to floral phase, environmental conditions and pollinator exposure. *Flora, 206*, 668-675.

Canto, A., & Herrera, C. M. (2012). Micro-organisms behind the pollination scenes: microbial imprint on floral nectar sugar variation in a tropical plant community. *Annals of Botany, 110*, 1173-1183.

Capriotti, A. (1953). I lievitti dei fiori. *Rivista di Biologia, 45*, 370–394.

Carter, C., & Thornburg, R. W. (2004). Is the nectar redox cycle a floral defense against microbial attack? *Trends in Plant Science, 9*, 320-324.

Carter, C., Shafir, S., Yehonatan, L., Palmer, R., & Thornburg, R. (2006). A novel role for proline in plant floral nectars. *Naturwissenschaften, 93*, 72-79.

Carter, C., Healy, R., O'Tool, N. M., Naqvi, S. M., Ren, G., Park, S., Beattie, G. A., Horner, H. T., & Thornburg, R. W. (2007). Tobacco nectaries express a novel NADPH oxidase implicated in the defense of floral reproductive tissues against microorganisms. *Plant Physiology, 143*, 389–399.

Chalcoff, V. R., Aizen, M. A., & Galetto, L. (2006). Nectar concentration and composition of 26 species from the temperate forest of South America. *Annals of Botany, 97*, 413–421.

Corbet, S. A., Unwin, D. M., & Prŷs-Jones, O. E. (1979). Humidity, nectar and insect visits to flowers, with special reference to *Crataegus, Tilia* and *Echium*. *Ecological Entomology, 4*, 9-22.

de la Barrera, E., & Nobel, P. (2004). Nectar: properties, floral aspects, and speculations on origin. *Trends in Plant Science, 9*, 65-69.

de Vega, C., Herrera, C. M., & Johnson, S. D. (2009). Yeasts in floral nectar of some South African plants: quantification and associations with pollinator type. *South African Journal of Botany, 75*, 798-806.

de Vega, C., & Herrera, C. M. (2012). Relationships among nectar-dwelling yeasts, flowers and ants: patterns and incidence on nectar traits. *Oikos, 121*, 1878-1888.

de Vega, C., Guzmán, B., Lachance, M. A., Steenhuisen, S. L., Johnson, S. D., & Herrera, C. M. (2012). *Metschnikowia proteae* sp nov., a nectarivorous insect-associated yeast species from Africa. *International Journal of Systematic and Evolutionary Microbiology, 62*, 2538-2545.

de Vega, C, & Herrera, C. M. (2013). Microorganisms transported by ants induce changes in floral nectar composition of an ant-pollinated plant. *American Journal of Botany, 100*, 792-800.

de Vega, C., Guzmán, B., Steenhuisen, S. L., Johnson, S. D., Herrera, C. M., & Lachance, M. A. (2014). *Metschnikowia drakensbergensis* sp. nov. and *Metschnikowia caudata* sp. nov., two endemic yeasts associated with *Protea* flowers in South Africa. *International Journal of Systematic and Evolutionary Microbiology*, doi: 10.1099/ijs.0.068445-0.

Dyer, A. G., Whitney, H. M., Arnold, S. E. J., Glover, B. J. & Chittka, L. (2006). Bees associate warmth with floral colour. *Nature, 442*, 525-525.

Ehlers, B., & Olesen, J. (1997). The fruit-wasp route to toxic nectar in *Epipactis* orchids. *Flora, 192*, 223-229.

Eisikowitch, D., Kevan, P., & Lachance, M. A. (1990). The nectar-inhabiting yeasts and their effect on pollen germination in common milkweed, *Asclepias syriaca* L. *Israel Journal of Botany, 39*, 217-225

Elisens, W. J., & Freeman, C. E. (1988). Floral nectar sugar composition and pollinator type among new world genera in tribe Antirrhineae (Scrophulariaceae). *American Journal of Botany, 75*, 971-978.

Fenster, C. B., Armbruster, W. S., Wilson, P., Dudash, M. R., & Thomson, J. D. (2004). Pollination syndromes and floral specialization. *Annual Review of Ecology Evolution and Systematics, 35*, 375-403.

Ferro Fontán, C., & Chirife, J. (1981). The evaluation of water-activity in aqueous-solutions from freezing-point depression. *Journal of Food Technology*, *16*, 21–30.

Fonseca, Á., & Inácio, J. (2006). Phylloplane yeasts. In: G. Péter, & C. A. Rosa, *Biodiversity and Ecophysiology of Yeasts*. Berlin, Springer-Verlag.

Galetto, L., Bernardello, G., & Sosa, C. A. (1998). The relationship between floral nectar composition and visitors in *Lycium* (Solanaceae) from Argentina and Chile: what does it reflect? *Flora, 193*, 303–314.

Galetto, L., & Bernardello, G. (2004). Floral nectaries, nectar production dynamics and chemical composition in six *Ipomoea* species (Convolvulaceae) in relation to pollinators. *Annals of Botany, 94*, 269-280.

Gardener, M. C., & Gillman, M. P. (2002). The taste of nectar: a neglected area of pollination ecology. *Oikos, 98*, 552–557.

Gilbert, L. E. (1972). Pollen feeding and reproductive biology of *Heliconius* butterflies. *Proceedings of the National Academy of Sciences of the United States of America, 69*, 1403–1407.

Golonka, A. M. (2002). Nectar-Inhabiting Micro-organisms and the Dioecious Plant Species *Silene latifolia*. PhD Thesis, Durham, NC, USA, Duke University.

Golonka, A. M., & Vilgalys, R. (2013). Nectar-inhabiting yeasts in Virginian populations of *Silene latifolia* (Caryophyllaceae) and coflowering species. *American Midland Naturalist, 69*, 235-258.

Golonka, A. M., Obi Johnson, B., Freeman, J., & Hinson, D. J. Impact of nectarivorous yeasts on *Silene caroliniana*'s scent. *Eastern Biologist, 3,* 1–27.

González-Teuber, M., & Heil, M. (2009). Nectar chemistry is tailored for both attraction of mutualists and protection from exploiters. *Plant Signaling and Behavior, 4*, 809-13.

Good, A. P., Gauthier, L.-P. L., Vannette, R. L., & Fukami, T. (2014). Honey bees avoid nectar colonized by three bacterial species, but not by a yeast species, isolated from the bee gut. *PLoS One, 9*, e86494.

Goodrich, K. R., Zjhra, M. L., Ley, C. A. & Raguso, R. A. (2006). When flowers smell fermented: The chemistry and ontogeny of yeasty floral scent in pawpaw (*Asimina triloba*: Annonaceae). *International Journal of Plant Sciences, 167*, 33-46.

Graham, J. R., Ellis, J. D., Benda, N. D., Kurtzman, C. P., & Boucias, D. G. (2011). *Kodamaea ohmeri* (Ascomycota: Saccharomycotina) presence in commercial *Bombus impatiens* Cresson and feral *Bombus pensylvanicus* DeGeer (Hymenoptera: Apidae) colonies. *Journal of Apicultural Research, 50*, 218-226.

Grüss, J. (1917). Die Anpassung eines Pilzes (*Anthomyces reukaufii*) an den Blütenbau und den Bienenrüssel. *Berichte der Deutschen Botanischen Gesellschaft, 35*, 746–761.

Hahn-Hägerdal, B., Karhumaa, K., Fonseca, C., Spencer-Martins, I., & Gorwa-Grauslund, M. F. (2007). Towards industrial pentose-fermenting yeast strains. *Applied Microbiology and Biotechnology, 74*, 937–953.

Halpern, M., Fridman, S., Atamna-Ismaeel, N., & Izhaki, I. (2013). *Rosenbergiella nectarea* gen. nov. sp. nov., in the family Enterobacteriaceae, isolated from floral nectar. *International Journal of Systematic and Evolutionary Microbiology, 63*, 4259–4265.

Heil, M. (2013). Nectar: generation, regulation and ecological functions. *Trends in Plant Science, 16*, 191-200.

Herrera, C. M. (2014). Population growth of the floricolous yeast *Metschnikowia reukaufii*: effects of nectar host, yeast genotype and host x genotype interaction. *FEMS Microbiology Ecology, 88*, 250-257.

Herrera, C. M., de Vega, C., Canto, A., & Pozo, M. I. (2009). Yeasts in floral nectar: a quantitative survey. *Annals of Botany, 103*, 1415-1423.

Herrera, C. M., Pérez, R., & Alonso, C. (2006). Extreme intraplant variation in nectar sugar composition in an insect-pollinated perennial herb. *American Journal of Botany*, *93*, 575-581.

Herrera, C. M., García, I. M., & Pérez, R. (2008). Invisible floral larcenies: Microbial communities degrade floral nectar of bumble bee-pollinated plants. *Ecology,* 89, 2369-2376.

Herrera, C. M., Canto, A., Pozo, M. I., & Bazaga, P. (2010). Inhospitable sweetness: nectar filtering of pollinator-borne inocula leads to impoverished, phylogenetically clustered yeast communities. *Proceedings of the Royal Society of London series B-Biological Sciences, 277*, 747-754.

Herrera, C. M., & Pozo, M. I. (2010). Nectar yeasts warm the flowers of a winter-blooming plant. *Proceedings of the Royal Society of London series B-Biological Sciences*, *277*, 1827-1834.

Herrera, C. M., Pozo, M. I., & Bazaga, P. (2012). Jack of all nectars, master of most: DNA methylation and the epigenetic basis of niche width in a flower-living yeast. *Molecular Ecology, 21*, 2602-2616.

Herrera, C. M., Pozo, M. I., & Medrano, M. (2013). Yeasts in nectar of an early-blooming herb: sought by bumblebees, detrimental to plant fecundity. *Ecology, 94*, 273-279.

Herrera, C. M., Pozo, M. I., & Bazaga, P. (2014). Nonrandom genotype distribution among floral hosts contributes to local and regional genetic diversity in the nectar living yeast *Metschnikowia reukaufii. FEMS Microbiology Ecology, 87*, 568-575.

Herrera, J. (1988). Aminoácidos en el néctar de plantas del sur de España. *Anales del Jardín Botánico de Madrid, 45*, 475-482.

Heslop-Harrison, J., & Heslop-Harrison, Y. (1988). Some permeability properties of angiosperm pollen grains, pollen tubes and generative cells. *Sexual Plant Reproduction, 1*, 65-73.

Jacquemyn, H., Lenaerts, M., Brys, R., Willems, K.A., Honnay, O., & Lievens, B. (2013a). Among-population variation in microbial community structure in the floral nectar of the bee-pollinated forest herb *Pulmonaria officinalis* L. *Plos One, 8*, e56917.

Jacquemyn, H., Lenaerts, M., Tyteca, D., & Lievens, B. (2013b). Microbial diversity in the floral nectar of seven *Epipactis* (Orchidaceae) species. *Microbiology Open, 2*, 644–658.

Johnson, M., Wratten, S. D., Landis, D. A., & Gurr, G. M. 2008. Recent advances in conservation biological control of arthropods by arthropods. *Biological Control, 45,* 172-175.

Kevan, P.G., Eisikowitch, D., Fowle, S., & Thomas K. (1988). Yeast-contaminated nectar and its effect on bee foraging. *Journal of Apicultural Research, 27*, 26-29.

Lachance, M. A., Starmer, W. T., Rosa C. A., Bowles J. M., Barker J. S. F., & Janzen D. H. (2001). Biogeography of the yeasts of ephemeral flowers and their insects. *FEMS Yeast Research, 1*, 1–8.

Lachance, M.A. (2006). Yeast biodiversity: how many and how much? In: C. A. Rosa, & G. Péter, *Biodiversity and Ecophysiology of Yeasts*. Berlin, Springer-Verlag.

Langenberger, M. W., & Davis, A. R. (2002). Temporal changes in floral nectar production, reabsorption and composition associated with dichogamy in annual caraway (*Carum carvi*; Apiaceae). *American Journal of Botany, 89*, 1588–1598.

Lee, F. J., Rusch, D. B., Stewart, F. J., Mattila, H. R., & Newton, I. L. G. (2014). Saccharide breakdown and fermentation by the honey bee gut microbiome. *Environmental Microbiology,* doi: 10.1111/1462-2920.

Lenaerts, M., Álvarez-Pérez, S., de Vega, C., Van Assche, A., Johnson, S. D., Willems, K. A., Herrera, C. M., Jacquemyn, H., & Lievens, B. (2014). *Rosenbergiella australoborealis* sp. nov., *Rosenbergiella collisarenosi* sp. nov. and *Rosenbergiella epipactidis* sp. nov., three novel bacterial species isolated from floral nectar. *Systematic and Applied Microbiology,* doi: 10.1016/j.syapm.2014.03.002.

Lievens, B., Hallsworth, J. E., Pozo, M. I., Ben Belgacem, Z., Stevenson, A., Willems, K. A., & Jacquemyn, H. (2014). Microbiology of sugar-rich environments: diversity, ecology, and system constraints. *Environmental Microbiology,* doi: 10.1111/1462-2920.12570.

Lin, Y. & Tanaka, S. (2006). Ethanol fermentation from biomass resources: current state and prospects. *Applied Microbiology and Biotechnology, 69,* 627-642.

Lin, I. W., Sosso, D., Chen, L.-Q., Gase, K., Kim, S.-G., Kessler, D., Klinkenberg, P., Gorder, M., Hou, B.-H., Qu, X.-Q., Carter, C., Baldwin, I., & Frommer, W. (2014). Nectar secretion requires sucrose phosphate synthases and the sugar transporter SWEET9. *Nature*, *508*, 546-549.

Majdak, A. Herjavec, S., Orlíc, S., Redžepović, S., & Mirošević. N. (2002). Comparison of wine aroma compounds produced by *Saccharomyces paradoxus* and *Saccharomyces cerevisiae* strains. *Food Technology and Biotechnology*, *40*, 103–109.

Manson, J. S., Lachance, M. A., & Thomson, J. D. (2007). *Candida gelsemii* sp nov., a yeast of the Metschnikowiaceae clade isolated from nectar of the poisonous *Carolina jessamine*. *Antonie Van Leeuwenhoek. International Journal of General and Molecular Microbiology*, *92*, 37-42.

Mares, D. (1987). Antimicrobial activity of protoanemonin, a lactone from ranunculaceous plants. *Mycopathologia, 98*, 133-140.

Mushtaq, M., Ayesha, J., & Sharfun, N. (2006). Biodiversity of yeast mycoflora in nectar of *Bombax cieba* and *Canna indica* flower. *Pakistan Journal of Botany, 38*, 1279-1288.

Mushtaq, M., Jamal, A., & Nahar, S. (2007). Biodiversity of yeast mycoflora in nectar of *Hibiscus rosa-sinensis* and *Ixora coccinea* flowers. *Pakistan Journal of Botany*, *39*, 1367-1376.

Mushtaq, M., Jamal, A., & Nahar, S. (2008). Biodiversity of yeast mycoflora in nectar of *Malvaviscus arboreus* and *Pancratium biflorum* flowers. *Pakistan Journal of Botany*, *40*, 877-885.

Nadson, G. A., & Krassilnikov, N. A. (1927). La levure du nectar des fleurs: *Anthomyces reukaufii* Gruess. *Bulletin de la Société Mycologique de France*, *43*, 232–244.

Napier, K. R, McWhorter, T. J., Nicolson, S. W., & Fleming, P. A. (2013). Sugar preferences of avian nectarivores are correlated with intestinal sucrase activity. *Physiological and Biochemical Zoology, 5*, 499 – 514.

Nepi, M. (2014). Beyond nectar sweetness: the hidden ecological role of non-protein amino acids in nectar. *Journal of Ecology, 102*, 108-115.

Nepi, M., Cresti, L., Guarnieri, M., & Pacini, E. (2011). Dynamics of nectar production and nectar homeostasis in male flowers of *Cucurbita pepo* L. *International Journal of Plant Sciences, 172*, 183-190.

Nicolson, S. W. (1994). *Eucalyptus* nectar: production, availability, composition and osmotic consequences for the larva of the eucalypt nectar fly, *Drosophila flavohirta. South African Journal of Science, 90*, 75–79.

Nicolson, S. W. (2002). Pollination by passerine birds: why are the nectars so dilute? *Comparative Biochemistry and Physiology - Part B: Biochemistry and Molecular Biology, 131*, 645–652.

Nicolson, S. W. (2007). Nectar consumers. In: S.W Nicolson, M. Nepi, & E. Pacini, *Nectaries and Nectar*. Dordrecht, Springer-Verlag.

Nicolson, S. W., de Veer, L., Köhler, A. & Pirk, C. W. W. (2013). Honeybees prefer warmer nectar and less viscous nectar, regardless of sugar concentration. *Proceedings of the Royal Society of London series B-Biological Sciences, 280*, 20131597.

Nicolson, S. W. & Thornburg, R. (2007). Nectar chemistry. In: S.W Nicolson, M. Nepi, & E. Pacini, *Nectaries and Nectar*. Dordrecht, Springer-Verlag.

Nicolson, S. W., & van Wyk, B.-E. (1998). Nectar sugars in Proteaceae: patterns and processes. *Australian Journal of Botany, 46*, 489–504.

Nocentini, D., Pacini, E., Guarnieri, M., Martelli, D., & Nepi, M. (2013). Intrapopulation heterogeneity in floral nectar attributes and foraging insects of an ecotonal Mediterranean species. *Plant Ecology, 214*, 799-809.

Ollerton, J., Winfree, R., & Tarrant, S. (2011). How many flowering plants are pollinated by animals? *Oikos, 120*, 321-326.

Ornelas, J. F., Ordano, M., De-Nova, A. J., Quintero, M. E., & Garland, T. (2007). Phylogenetic analysis of interspecific variation in nectar of hummingbird-visited plants. *Journal of Evolutionary Biology, 20*, 1904–1917.

Park, S., & Thornburg, R. (2009). Biochemistry of nectar proteins. *Journal of Plant Biology, 52*, 27-34.

Peay, K., Beslisle, M., & Fukami, T. (2012). Phylogenetic relatedness predicts priority effects in nectar yeast communities. *Proceedings of the Royal Society of London series B-Biological Sciences, 279*, 749–758.

Percival, M. S. (1961). Types of nectar in angiosperms. *New Phytologist, 60*, 235–281.

Petanidou, T. (2005). Sugars in Mediterranean floral nectars: an ecological and evolutionary approach. *Journal of Chemical Ecology, 31*, 1065–1088.

Petanidou, T., Van Laere, A. J.., & Smets, E. (1996). Change in floral nectar components from fresh to senescent flowers of *Capparis spinosa* (Capparidaceae), a nocturnally flowering Mediterranean shrub. *Plant Systematics and Evolution,* 199, 79-92.

Petanidou, T., Van Laere, A. N., Ellis, W., & Smets, E. (2006). What shapes amino acid and sugar composition in Mediterranean floral nectars? *Oikos, 115*, 155-169.

Pozo, M. I. (2012). Yeasts in Floral Nectar: Community Ecology and Interactions with Insect Pollinators and Host Plants. Unpublished D. Phil. Thesis, Seville, Spain, University of Seville.

Pozo, M. I., de Vega, C., Canto, A., & Herrera, C. M. (2009). Presence of yeasts in floral nectar is consistent with the hypothesis of microbial-mediated signaling in plant-pollinator interactions. *Plant Signaling and Behavior, 4*, 1102-1104.

Pozo, M. I., Herrera, C. M., & Bazaga, P. (2011). Species richness of yeast communities in floral nectar of southern Spanish plants. *Microbial Ecology, 61*, 82-91.

Pozo, M. I., Herrera, C. M., & Alonso, C. (2014). Spatial and temporal distribution patterns of nectar-inhabiting yeasts: how different floral microenvironments arise in winter-blooming *Helleborus foetidus. Fungal Ecology, 11,* 173-180.

Pozo, M. I., Lachance, M. A., & Herrera, C. M. (2012). Nectar yeasts of two southern Spanish plants: the roles of immigration and physiological traits in community assembly. *FEMS Microbiology Ecology, 80*, 281-293.

Raguso, R. A. (2004). Why are some floral nectars scented? *Ecology, 85*, 1486-1494.

Raguso, R. A. (2008). Wake up and smell the roses: the ecology and evolution of floral scent. *Annual Review of Ecology, Evolution, and Systematics, 1*, 549-569.

Rands, S. A. & Whitney, H. M. (2008). Floral temperature and optimal foraging: is heat a feasible floral reward for pollinators? *PLoS One, 3*, e2007.

Rosa, C. A., Lachance, M. A., Silva, J. O. C., Teixeira, A. C. P., Marini, M. M., Antonini, Y., & Martins, R. P. (2003). Yeast communities associated with stingless bees. *FEMS Yeast Research, 4*, 271-275.

Roughgarden, J. (1972). Evolution of niche width. *American Naturalist, 106*, 683–718.

Sandhu, D., & Waraich, M. (1985). Yeasts associated with pollinating bees and flower nectar. *Microbial Ecology, 11*, 51-58.

Schaeffer, R. N., & Irwin, R. E. (2014). Yeasts in nectar enhance male fitness in a montane perennial herb. *Ecology, 95*, 1792–1798.

Schoellhorn, K. (1919). Sur la fermentation de quelques levures des nectars des plantes d'hiver. *Bulletin de la Société botanique de Genève, 11*, 154–190.

Schuster, V., & Úlehla, V. (1913). Studien über Nektarorganismen. *Berichte der Deutschen Botanischen Gesellschaft, 31*, 129–139.

Sereno, A. M., Hubinger, M. D., Comesaña, J. F., & Correa, A. (2001). Prediction of water activity of osmotic solutions. *Journal of Food Engineering, 49*, 103-114.

Starzak, M., & Mathlouthi, M. (2006). Temperature dependence of water activity in aqueous solutions of sucrose. *Food Chemistry, 96*, 346-370.

Sultan, S. E. & Spencer, H. G. (2002). Metapopulation structure favors plasticity over local adaptation. *American Naturalist, 160*, 271–283.

Swiegers, J. H., Bartowsky, E. J., Henschke, P. A., & Pretorius, I. S. (2005). Yeast and bacterial modulation of wine aroma and flavour. *Australian Journal of Grape and Wine Research, 11*, 139-173.

Tucker, C. M., & Fukami, T. (2014). Environmental variability counteracts priority effects to facilitate species coexistence: evidence from nectar microbes. *Proceedings of the Royal Society of London series B-Biological Sciences, 281*, 20132637.

van Wyk, B. E., Whitehead, C. S., Glen, H. F., Hardy, D. S., van Jaarsveld, E. J., & Smith, G. F. (1993). Nectar sugar composition in the subfamily Alooideae (Asphodelaceae). *Biochemical Systematics and Ecology, 21*, 249–253.

Vannette, R. L., Gauthier, M. P. L., & Fukami, T. (2013). Nectar bacteria, but not yeast, weaken a plant-pollinator mutualism. *Proceedings of the Royal Society of London series B-Biological Sciences, 280*, 20122601.

Vörös-Felkai, G. (1957). Données sur les levures de fleurs répandues en Hongrie. *Acta Botanica Academiae Scientiarum Hungaricae, 3*, 391–399.

Weiss, M. (2001). Vision and learning in some neglected pollinators. In: L. Chittka, & J. D. Thomson, Cognitive Ecology of Pollination, Animal Behavior and Floral Evolution. Cambridge, UK, Cambridge University Press.

Whitney, H. M., Dyer, A., Chittka, L., Rands, S. A. & Glover, B. J. (2008). The interaction of temperature and sucrose concentration on foraging preferences in bumblebees. *Naturwissenschaften, 95*, 845-850.

Wiens, F., Zitzmann, A., Lachance, M. A., Yegles, M., Pragst, F., Wurst, F. M., von Holst, D., Guan, S. L. & Spanagel, R. (2008). Chronic intake of fermented floral nectar by wild treeshrews. *Proceedings of the National Academy of Sciences of the United States of America, 105*, 10426-10431.

Zender, G., Gurr, G.M., Kuhne, S., Wade, M. R., Wratten, S. D., & Wyss, E. (2007). Arthropod pest management in organic crops. *Annual Review of Entomology, 52*, 57-80.

In: Nectar
Editor: Robin L. Peck

ISBN: 978-1-63463-650-6

Chapter 2

EFFECT OF NECTAR COMPOSITION ON BEE ATTRACTION FOR ONION SEED PRODUCTION

Verónica Soto[1], María Fernanda Silva[1] and Claudio Galmarini[2,3,4]***

[1]Instituto de Biología Agrícola de Mendoza (IBAM-CONICET), Chacras de Coria, Mendoza, Argentina

[2]Instituto Nacional de Tecnología Agropecuaria (INTA), Estación Experimental Agropecuaria La Consulta, La Consulta, Mendoza, Argentina

[3]Consejo Nacional de Investigaciones Científicas y Técnicas (CONICET), Argentina

[4]Facultad de Ciencias Agrarias, Universidad Nacional de Cuyo, Chacras de Coria, Luján, Mendoza, Argentina

ABSTRACT

Onion is the second most cultivated vegetable crop in the world. Two types of onion varieties are usually grown around the world; open pollinated (OP) and first generation (F1) hybrids. Honey bees are the main pollinators of onion seed production, but owing to low

* Corresponding author email: galmarini.claudio@inta.gob.ar Fax: +54-261-4960469; Tel: +54-261-4964208.

attractiveness of flowers, pollination is often inadequate. Pollination problems result in low seed yields. Field observations indicate that F1 hybrid seed yields are much lower than open-pollinated varieties seed yields, with a decrease of up to 60%. Nectar is the most important floral reward offered by plants. Floral nectar composition has been studied as the potential cause of bee preferences in many species, among them onion. Nectar usually does not repel bees, but some nectar may be less attractive than nectar of competing flowers. Honeybees have distaste for onion flowers and avoid visiting them. Frequently, onion nectar potassium concentration and nectar sucrose concentration have been mentioned as potential causes of bee unattractiveness. Also, minor components of nectar such as phenols, amino acids, proteins, organosulfur compounds and alkaloids might directly affect bee foraging. There are differences in the chemical composition of nectar and floral morphology between open pollinated and F1 Hybrids and also within F1 Hybrids. These differences are correlated with the number of visits and seed yield. Therefore, it would be possible to select indirectly the most promising seed production lines analyzing nectar or floral morphological characters.

INTRODUCTION

Onion (*Allium cepa* L.) a member of family *Alliaceae* is one of the most important vegetable crops grown worldwide (Abrol 2010). Its fleshy bulbs are used both in fresh and dehydrated form (Pathak 1999). Seeds are the usual way that onion can be reproduced. Onion follows a growth cycle which is normally biennial: the edible bulb is produced in the first growing season. Left to itself or replanted after storage, the onion blooms in the second season of growth, utilizing the food reserves in the bulbs to produce substantial flower stalks and umbels with hundreds or even thousands of small flowers (Currah 1990). The individual flowers of this species are hermaphrodite and fertile but exhibit protandry, that is, the anthers release pollen before the stigma is receptive (Acosta et al. 1993). Nevertheless, protandry does not prevent flowers of the same umbel to be selfed (Currah and Ockendon 1978).

Cross pollination is the most frequently used method for seed production, which depends heavily on insects for any significant increase in seed yield. Although numerous insect species have been reported to visit onion flowers for nectar and pollen, honeybees (*Apis mellifera* L.) have been reported as the most efficient and practical pollinator of this crop because of their affinity for pollen and nectar and efficiency in pollen transfer (Abrol 2010). Nevertheless, widespread use of the honey bee as pollinators not always brings the expected results because the onion nectar is not particularly attractive (Silva et al. 2004).

Onion is an allogamous specie, which allows the exploitation of heterosis, and hybrids are increasingly used. Open-pollinated (OP) and first generation (F1) onion hybrids are usually grown around the world. Cytoplasmic-genic male sterility (CMS) systems are used to produce hybrid-onion seed (Brewster 1994). To produce F1 hybrid seed in onion, it is necessary to cross a male sterile line (MSL) with a fertile one. F1 hybrids have the advantage of greater uniformity and high yields in the production of bulbs; however, seed production is inconsistent, producing less seeds than OP cultivars. As a consequence, to be economically viable, they must have a good seed yield, otherwise F1 hybrids become too expensive for growers. Moreover, highly variable seed yields between different onion hybrids have been observed (Silva and Dean 2000). Part of the problem is attributable to the relatively low attractiveness of some male sterile lines to insects (Parker 1982). Cross-pollination of male and female genetic lines also depends on honey bees (Silva et al. 2004). To achieve efficient transfer of pollen to the male sterile rows, a numerous population of pollinators is necessary, and if possible the male sterile lines should offer as much nectar as the male fertile ones. Only honeybees are sufficiently transportable and available for hire to onion seed producers (Currah 1990). Consequently, seed yield is closely correlated with the behavior of honey bees in onion seed fields (Benedek 1976).

Factors Affecting Onion Pollination

Flower attributes such as color, size, and specially nectar volume and composition are considered important factors for honey bee attraction. Minor components of nectar may directly affect bee foraging, e.g., phenols, amino acids, reducing acids, lipids, proteins, and alkaloids (Torres and Galetto 1998, Butler 1945), and among *Allium* species some features may not be attractive to pollinators, such as organosulfur compounds or unusually high levels of potassium, which might deter pollinators (Silva et al. 2003, Soto et al. 2013). Therefore, low pollination further accentuates the necessity for biological data on onion nectar.

Floral traits: This factors act as components of a "floral integrated design", and therefore their size and variability can be related to the pollination process, i.e., they cannot vary independently (Fernández et al. 2009). The onion umbel is a roughly spherical inflorescence composed of a numerous open, dish-shaped flowers with easily accessible nectar. The umbel presents a substantial landing surface composed of the expanded, white, green-veined tepals or perianth segments, and the upper surfaces of the ovaries. Onions may have less than 100 to over 2000 flowers per umbel, largely depending on the bulb size and vigor and the numbers of umbels produced by each bulb. In each flower, nectar is exposed in three small wells or cups, formed between the lower ovary wall where the nectarines are located underneath a small flap or tissue, and the broad filament bases of the inner whorl of three stamens (Rabinowitch and Brewster 1990) (Figure 1).

Havey (1999) reported morphological differences between male sterile and fertile flowers of *Allium galanthum*; male-sterile flowers have a curved perianth and filaments of stamens do not have anthers. These features hinder access to nectar and reduce visits of pollinating bees with a concomitant decrease of seed yield. Nevertheless, studies about the influence of floral traits on onion pollination and the differences between hybrid and open pollinated, have not been deeply studied yet. A first study in three male sterile lines was reported by Soto et al. (2013) and they found a negative correlation between style length with bee preference and seed yield, while a positive relationship was found between tepal and stamen length with seed yield and bee foraging behavior.

On the other hand, most onion pollination research is focused on the kind of insects that pollinated umbels, the frequency of visits, the effect of pedoclimatic factors or insecticide use on pollinators (Benedek 1976, Brown et al. 1977, Parker 1982, Currah and Ockendon 1984, Saeed et al. 2003, Silva et al. 2004, Wilkaniec et al. 2004, Sajjad et al. 2008, Abrol 2010, Long and Morandin 2011). Little data are available concerning nectar properties as they relate to honey bee visitation and subsequent seed production in agricultural crops, especially in onion (Benedek 1976, Hagler 1990, Silva and Dean 2000, Sajjad et al. 2008, Abrol 2010, Soto et al. 2013). Nectar usually does not repel bees, but a particular nectar may be less attractive than nectar of competing flowers (Afik et al. 2006). Honey bees have distaste for flowers of certain varieties of onion and avoid visiting them, and the appearance of different sources of nourishment can easily pull away these bees from flowering crops (Silva et al. 2004). Honey bees forage selectively among onion cultivars and are sensitive to qualitative and quantitative differences in floral cues and rewards. Development of hybrid onions may have altered flower nectar and aroma chemistry. Orians in 2000, expressed that even though plant breeders use hybridization to obtain or introgress desired traits, surprisingly little attention has been paid to the effects of hybridization on secondary chemistry and the survival of hybrids in natural populations. Clearly, secondary chemistry may be an important component of hybrid fitness and influence speciation, introgression, and plant–herbivore interactions. Loss of attractiveness usually results in reduced pollinator visitation, reduced pollen transport, and reduced seed yields (Hagler et al. 1990). Hence, some authors suggest that onion MSL flowers produce less nectar, as a consequence recommend doubling the number of pollinators for onion hybrid seed production (Wilkaniec et al. 2004).

The lack of pollination of onion crops have been mainly attributed to some onion nectar characteristics such as potassium concentration and sugar composition. Theoretically, each of these components could affect honey bee visits. Nevertheless, these studies were conflicting or inconclusive (Silva and Dean 2000). An example of onion nectar composition of fertile and male sterile lines is shown in Table 1.

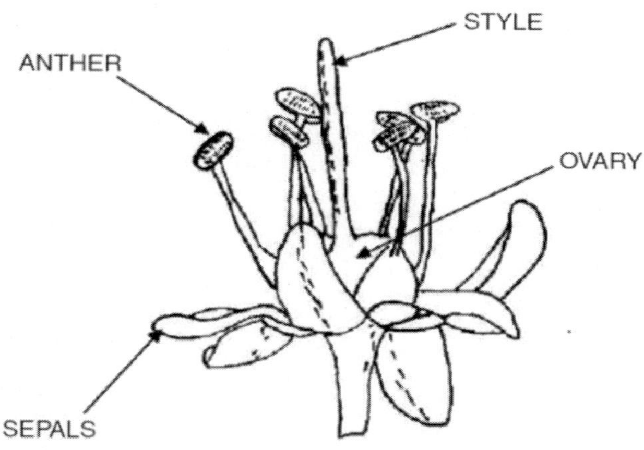

Figure 1. Onion flower (Brewster 2008).

Potassium concentration: Before 2000, nearly all previously reported research had attributed to the exceptionally high potassium concentrations in the nectar of flowers (up to 13,000 ppm), the lack of pollination of onion crops (Waller et al. 1972, Waller et al. 1976, Hagler 1990). These authors determined that onion nectar is rich in potassium content and concluded that these high potassium levels are the primary factor limiting successful bee pollination of onions and resulting in poor pollination of onion crops. Then, in 2000, Silva and Dean reported that nectar potassium concentration was not correlated with bee pollination. In addition, these authors reported a great variation in nectar potassium concentration among different hybrid parents. In agreement with these authors, Soto et al., 2013 reported that in spite of the high potassium concentration found in all studied onion nectars; this factor has no relationship with the onion fitness of F1 hybrids.

Table 1. Onion nectar composition of fertile and male sterile lines (Soto et al. 2013)

	Line			
Trait	OP	MSL 1	MSL 2	MSL 3
Glucosa conc.	54.53 ± 0.325 a	25.10 ± 0.95 d	38.53 ± 0.65 b	27.46 ± 0.64 c
Fructosa conc.	78.73 ± 1.4 a	45.50 ± 2.52 c	66.26 ± 2.17 b	47.36 ± 1.01 c
Total Sugar	133.26 a	70.60 c	104.80 b	74.83 c
Potassium conc.	3230988.5 ± 59373.5 b	3330107 ± 11461.5 ab	3358535 ± 57874.0 a	2790578.7 ± 32568.9 c
Magnesium conc.	124279 ± 2765.5 a	154333 ± 3310.0 b	153555.5 ± 831.0 b	157443 ± 3148.9 b
VOCs conc.	168.96 ± 11.91 b	247.14 ± 24.12 a	192.95 ± 18.42 b	153.19 ± 4.96 b
VOCs repellent conc.	23.48 ± 3.28 b	38.16 ± 0.65 a	42.02 ± 5.14 a	23.53 ± 1.32 b
Vinylphenol	nd	1.17 ± 0.03 a	1.11 ± 0.11 a	nd
Catechin	nd	nd	nd	2.57 ± 0.11
Naringenin	nd	8.01 ± 1.28 a	2.57 ± 0.47 b	2.52 ± 0.40 b
Cinnamic acid	nd	nd	0.57 ± 0.01	nd
Chlorogenic acid	nd	nd	2.14 ± 0.10 a	1.84 ± 0.08 b
Quercetin	nd	1.04 ± 0.07 a	0.19 ± 0.009 b	nd
Vanillic acid	0.81 ± 0.03 a	0.82 ± 0.03 a	nd	nd
Luteolin	7.20 ± 0.23 a	nd	nd	0.67 ± 0.03 b
Caffeic acid	1.10 ± 0.01 a	1.10 ± 0.01 a	nd	0.76 ± 0.03 b
TPC	9.12 ± 0.19 b	11.78 ± 1.91 a	6.52 ± 0.31 c	8.18 ± 0.15 bc

OP: open pollinated cultivar. MSL: male sterile line. conc.: concentration. nd: not detected. VOCs: volatile organic compounds. TPC: total phenol content. Values represent mean ± SD of 3 determinations. Values in the same column with different letters present significant differences $P < 0.05$. Sugar concentration is expressed as g/l of nectar. Mineral composition is expressed as µg/l of nectar. Phenolic content is expressed as mg/l of nectar.

Sugar levels: The question of the relationship of sugar levels and onion nectar composition to honey bee foraging has not been resolved totally yet, in part due to limited nectar amounts, which difficult to conduct preference experiments based on manually collected nectar (Afik et al. 2006). In other species studies had shown that honey bees prefer a sucrose-rich liquid, but if this is not available, bees will collect nectar that is hexose-dominant (Nicolson 2011).

Waller et al. in 1972 did bioassays to evaluate honey bee response to sugar solutions with different fructose, glucose, and sucrose ratios, demonstrating that mixtures containing equal parts of fructose and glucose were attractive to honey bees. Later, Waller in 1974 reported that neither the sucrose/hexose ratio, nor any other single aspect of nectar sugar composition appears to exclusively determine the foraging behavior of these pollinators. The 3 predominant sugars in nectar are sucrose, glucose, and fructose, all of which are acceptable to bees. Onion nectar is hexose dominant with 53-56% fructose, 40-43% glucose, and only 4% sucrose. Thus, although sugar ratios in onion nectar are not optimal for bees, the results of Waller suggest that these ratios do not inhibit the collection of onion nectar.

In 1990, Hagler studied six different onion cultivars and reported that onion flowers provide enough high-quality nectar that, theoretically, will assure abundantly foraging honey bees visits. All onion cultivars tested had similar amounts of fructose and glucose (about 1:1 fructose/glucose ratios). Sucrose was the least abundant sugar. The ratio of sugars in onion nectar, therefore, should not inhibit honey bees from collecting it. Nevertheless, the concentration of sugars in onion nectar vary from 52 to 65% and these high sugar concentrations might deter honey bee foraging because they are too viscous for the honey bee to imbibe easily. Then, in another study Hagler et al. (1990), found that in simulated nectars which had an approximately equal concentration of potassium, bees preferred nectar with a high carbohydrate content. Other authors, Kumar and Kumar Gupta in 1993, studied the relationship between nectar sugar production with pollinators of onion and other Alliums. They showed that honeybees preferred the nectar of *A cepa*. Among honeybees, *Apis dorsata* dominated the blooms as compared to *A mellifera* and *A cerana indica*. These authors said that the differential preference of *A cepa* by honeybees might be associated with the concentration of nectar solutes coupled with the amount of dry nectar sugars.

On the other hand, authors who studied specifically hybrid onions obtained different results; Silva and Dean in 2000, reported that nectar sugar concentrations of the nectar of nine hybrid onion parent genotypes did not significantly correlate with the number of bee visits received by an umbel. Instead, the amount of nectar produced by both the umbels and the individual flowers was correlated with the number of bee visits. Nevertheless, Soto et al. 2013 reported that nectar sugars were consistently related to bee visits and seed yield. Moreover, the correlation with bee visits was higher for glucose than fructose.

Volatile compounds: The set of volatile organic compounds (VOCs) present in nectar comes both from the volatiles released by the surrounding floral tissues and from those released by the nectaries into the nectar solution. The function of those VOCs associated with the presence of nectar is probably complex, given that these scents have both attractive and deterrent effects on nectar consumers (Farré-Armengol et al. 2013). As mentioned above, there is a lack of knowledge about onion nectar composition and especially about volatile compounds of F1 hybrid onions and their effect on pollination. Among the studies of volatile compounds in onion, most of them were made in bulbs or in the plant.

Volatile sulfur-containing flavor compounds are responsible for the characteristic smell and taste of members of the *Alliaceae*. These compounds are alk(en)yl-thiosulfinates formed by the action of alliinase on odorless and nonvolatile cysteine-derived substances (*S*-alk(en)yl-L-cysteine sulfoxides (ACSOs), when these reactants are released by plant material disruption (Brewster 1994). Briggs and Goldman (2002), reported that sulphur compounds may have different purposes in flowers; umbels might need additional reserves of nitrogen and sulphur for pollen production, synthesis of pollinator attractants and seed production. As a defense against pests, mobilization of ACSOs could improve survival and fitness of the onion plant. During vernalization, the bulb would be protected by the ACSOs. Moreover, organosulphur compounds would be useful in warding off animals that would attack newly emerging fleshy tissues such as leaves and scapes, as well as high pungency in seedling sprouts and umbels containing developing seeds would improve progeny survival (Briggs and Goldman 2002).

Mochizuki et al. (1998) reported variation among onion plants in their volatile sulfur compounds and their compositions, and described that 1-propenyl and methyl groups are commonly found in onion, scallion, shallot, leek, and chive (Mochizuki et al. 1998). In 2003 Silva et al., designed a conditioning experiment in which odors actually identified from the onion flower were used, with an end goal of attracting greater numbers of foraging bees to the onion crop versus two competing crops (carrot and alfalfa). When fed the onion-scented sucrose solution, the degree of this preference declined compared with the other crops. The lack of increase in the level of foraging on the onion flowers may have been caused by the nature of the scent mixtures. Although most of the major onion floral scents were present, the scented solutions were not an exact replication of the actual onion floral odor. Dipropyl disulfide, the compound very characteristic of onion odor, is unique to *Allium* flowers.

Later, Storsberg et al. (2004) studied *Allium* species, *A. cepa* L. and *A. kermesinum* Rchb. as well as the resulting hybrids, they reported that among volatile sulfur-containing substances the main compounds of *A.cepa* were propyl mercaptan, dimethyl thiophene, dipropyl disulfide, and the two isomeric forms of propyl 1-propenyl disulfide. *A. kermesinum* showed 3 times more different volatile sulfur-containing metabolites. Where, the leading compounds were allyl mercaptan, allyl methyl sulfide, allyl methyldisulfide, and diallyl disulfide. As expected, with regard to the cysteine sulfoxide pattern, the hybrid plant showed the highest number of sulfurous volatiles, with the leading compounds allyl propyl disulfide and diallyl disulfide. Every compound detected in the profile of the parent plants was also present in the profile of *A. cepa* x *A. kermesinum*.

In a recent study, Soto et al., (unpublished) reported the relationship between volatile organic compounds and bee pollination and seed yield in different onion lines, including F1 hybrids. Among the 96 compounds identified, the sulfur compounds were the most common chemical class and differences in volatile sulfur compounds composition among examined nectar of onion lines were found. Methyl and propyl groups were the dominant compounds observed in onion hybrids. Sulfides, including dimethyl disulfide, dimethyl trisulfide, dimethyl tetrasulfide and methylthio sulfides isolated from studied samples arise from the degradation of amino acids containing sulfur.

Alkyl-sulfide compounds as well as alkyl-substituted piperidines and dioxolanes were found in at higher amounts in the male sterile lines (MSL) than the open pollinated line (OP), derivates of these compounds could act as insect repellents, even as insecticide (Xue et al. 2001, Scott et al. 2005, Chung et al. 2011, Murali and Chandrasekhar 2012, Dawid et al. 2012). Indeed, a negative relationship between the amount of "*repellent*" compounds and bee pollination was found. Besides, those lines which showed greater amounts of organosulfur compounds had lower bee pollination and seed yield. It was found that both, floral and nectar VOCs of F1 hybrids had an influence on bee attraction. A direct negative relationship between the volatile profile and pollinator foraging behavior for each onion line was observed. Honeybee preference trials suggest that open pollinated variety is more attractive than F1 hybrids. Nevertheless, the differences found on the volatile profile among MSLs were markedly related to seed yield. Authors said that the determination of VOCs can be use by breeders as a tool for the onion lines selection with higher potential to attract bees, and as a consequence, produce more seeds.

Phenolic content: Although the presence of secondary compounds in floral nectar may provide reproductive benefits to plants in some special cases, it is predicted to have detrimental effects if pollinators are deterred from visiting flowers. Paradoxically, floral nectar of some plant species also contains secondary compounds such as phenols that occur in leaves, stems, and roots to defend against attack by herbivores and microorganisms (Gegear et al. 2007). Phenolic substances are quite widespread in nectars. Their accumulation may turn the nectar toxic, becoming repellent to some visitors (Torres and Galetto 1998). Moreover, phenolic substances are also relatively common scent compounds of flowers. As well as attracting pollinators or repelling nectar thieves, these scent compounds may have a defensive function (Nicolson and Thornburg 2007).

On the other hand, it is well known that onion is one of the richest sources of dietary flavonoids in many countries (Mogren et al. 2006, Corzo-Martínez et al. 2007, Pérez-Gregorio et al. 2011). In spite of this, small literature base is available on the phenolics present in in onion nectar samples.

Phenolic compounds have been widely studied in honey (Ferreres et al. 1996). Compounds such as quercetin, kaempferol, luteolin, naringenin, cinnamic, vanillic, or caffeic acid have been reported as the most important phenols found in honey, pollen, propolis and other bee products related to floral nectar (Gegear et al. 2007, Nicolson and Thornburg 2007). Among the studies focused on nectar, acacia (Gil et al. 1995), and rosemary (Truchado et al. 2008) have been analyzed with the objective to find biochemical markers. Owing to it is difficult to obtain sufficient amounts of floral nectar necessary for behavioural studies and, honeybees gather large amounts of nectar and store it in the hive in a concentrated form as honey, maybe is the reason why nearly all of studies have been made on honey.

There were no reports on the phenolic content in onion nectar, on MSLs and their effect on bee preference and seed yield until a recent study published in 2014 by Soto et al. Their results showed markedly qualitative and quantitative analytical differences in the phenolic profile of male sterile and open pollinated lines. Nine different phenolic compounds were found in the studied samples. One male sterile line, have the nectar with the highest phenolic content. A compound such as naringenin was only found in male sterile lines while the open pollinated (OP) line showed the highest amount of luteolin. Moreover, most of the phenolic compounds found in our study are in agreement with those reported in honey; these could be explained due to their solubility in aqueous solutions (Raguso 2004). Then, according with the results obtained, it can be assumed that compounds such as naringenin and luteolin may influence bee foraging behavior, as deterrent and attracting agents, respectively.

Alkaloid compounds: These compounds have been studied extensively either in other species (Kretschmar and Baumann 1999, Gegear et al. 2007, Adler and Irwin 2012) or in preference test (Masters 1991, Afik et al. 2008, Köhler et al. 2012). Among these, nicotin and caffeine have been the most studied alkaloids. Köhler et al. (2012) reported that the feeding response to nectar secondary metabolites is dose-dependent: low concentrations of some phenolics and alkaloids are preferred by honeybees to sugar-only controls, while high concentrations inhibit ingestion. The attractive or deterrent effect of these secondary metabolites may depend not only on the concentration of the compound in question but also on the sugar concentration of the nectar. Nonetheless, there are not data of alkaloid content in onion nectar.

Aminoacids and proteins: *Allium* spp. characteristically contain a high concentration of non-protein sulphur amino acids (between 1 and 5% dry wt). Cysteine and methionine are the largest sulfur- containing amino acids in onion (Lancaster and Shaw 1989). Among these secondary metabolites, the S-alk(en)yl-L-cysteine sulphoxides are the most important. In the intact cell the sulphoxides are located in the cytoplasm and the hydrolytic enzyme alliinase in the vacuole. Disruption of the cell results in hydrolysis of the sulphoxides to volatile sulphides. Four sulphoxides are found to occur in *Allium* spp.: (+)-S-methyl-I_-cysteine sulphoxide (MCSO), (+)-S-propyl-L-cysteine sulphoxide (PCSO), trans-(+)- S-(1-propenyl)-L-cysteine sulphoxide (1-PECSO)and (+)-S-(2-propenyl)-L-cysteine sulphoxide, (2-PECSO).

The quantitative and qualitative differences in flavor precursor content result in the different flavours of many of the *Allium* spp., particularly the presence or absence of 1-PECSO which is responsible for the lachrymatory effect of onions and 2-PECSO which produces the characteristic taste of garlic (Brewster 2008).

Allium spp. also contains other non-volatile sulphur compounds, the y-glutamyl peptides, which are not acted upon by alliinase and do not give rise to flavour. In *Allium* spp. 18 y-glutamyl peptides of sulphur compounds have been identified. The significance of the y-glutamyl peptides in the metabolism of the plant has been unclear. They have generally been considered to function as reserves of nitrogen and sulphur (Lancaster and Shaw 1989). In spite of the vast research made in bulbs or in other parts of the plant, there is not information regarding nectar organosulfur composition and its relationship with bee visits or seed yield.

Correlations between nectar chemical composition, floral traits, pollination, and seed production:

It has been reported that frequency of honey bee visits and seed yield were highly related (Figure 2) (Silva and Dean 2000, Soto et al. 2013); a great difference between seed yield of the hybrids and OP cultivars were reported, these cultivars have higher amount of seed than the F1 hybrid. Also differences in seed yield among the hybrids were found.

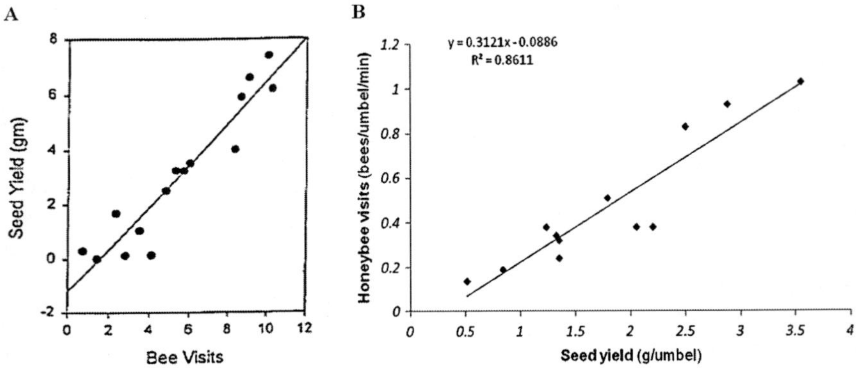

Figure 2. Relationship between seed yield and the number of honey bee visits (A): Silva and Dean 2000 (B): Soto et al., 2013.

Silva and Dean (2000) reported that nectar volume and nectar variability were most consistently related to bee visits and seed yield. Moreover, no significant correlation was found between nectar potassium concentration and the attractiveness of hybrid parents (1996, r2 5 0.03; P , 0.15; 1997, r2 5 0.26; P , 0.64). On the other hand, Soto et al. (2013) reported that there is a clear relation between the flower morphology, nectar composition, and honey bee behavior for each onion line. Among the morphological characteristics, the size of the nectary, anthers, and tepals play a key role in attracting bees. Within the analytical characterization, they found compounds that could act as attractants of bees, such as sugars, luteolin, or vanillic acid. Other factors may be acting negatively in pollinator attraction, such as style length; naringenin and vinylphenol concentration; volatile compounds such as dioxolanes, piperidines, and organosulfur compounds; and some metals like Mg, which were found in greater amounts in MSLs (Table 2).

Table 2. Correlations between nectar chemical composition, floral traits, frequency of bee visits, and seed yield (Soto et al. 2013)

Trait	Bee visits	Seed yield
Outer anther area	0.854***	0.662*
Intern anther area	0.881***	0.725**
Outer tepal area	0.941***	0.879**
Intern tepal length	0.961***	0.881**
Ovary area	0.070ns	0.090ns
Style length	−0.916***	−0.854**
Nectary vol	0.983***	0.937***
Fructose concn	0.882***	0.925***
Glucose concn	0.951***	0.939***
Potassium concn	0.038ns	0.138ns
Magnesium concn	−0.911***	−0.798**
VOCs repellents concn	−0.661*	−0.759**
Naringenin concn	−0.786**	−0.815**
Luteolin concn	0.798**	0.936*
Bee visits		0.928***

Concn, concentration; vol, volume; ns, not significant.
The symbols *, **, and *** denote significance at $P < 0.05, 0.01$, and 0.001, respectively.

Conclusion

Few studies have been done on the composition of onion nectar and its relationship with bee foraging behavior; most of these works have focused on sugar and potassium concentration. However, so far little is known about the metabolites that are commonly found in onion nectar and the overall characterization of nectar and floral morphology of onion have not been deeply studied.

A positive correlation is found between bee visits and seed yield. Among all research reported related to onion nectar composition, positive and negative correlations with the number of bee visits and seed yield has been found. Nectar sugars are the most consistently related to both, bee visits and seed yield. The correlation with bee visits was higher for glucose than fructose. Within the analytical characterization, it was found that some compounds could act as attractants of bees, while other factors could act negatively in pollinator attraction

F1 hybrids normally produce lees seeds than OP cultivars. Moreover, highly variable seed yields between different onion hybrids have been observed. Therefore, further studies should be done regarding onion nectar composition of MSLs and their effect on bee preference and seed yield. Without a very significant increase in foraging behavior on the target crop, the economics would not prove to be beneficial to seed growers. Instead of trying to train honey bees to forage on a crop that is not energetically rewarding, it would be better for plant breeders to improve the food source itself. It has been shown that onions do not provide enough nectar to make the crop profitable and thus attractive to honey bees. Therefore, improving the reward should be the ultimate goal in achieving a permanent solution to the problem of poor pollinator activity.

REFERENCES

Abrol, D. P. (2010) Foraging behaviour of *Apis florea* F., an important pollinator of *Allium cepa* L. *Journal of Apicultural Research,* 49, 318-325.

Acosta, A., J. C. Gaviola & C. Galmarini (1993) Manual de producción de semillas hortícolas: Cebolla. *Asociación Cooperadora INTA La Consulta*, 1-83.

Adler, L. S. & R. E. Irwin (2012) Nectar alkaloids decrease pollination and female reproduction in a native plant. *Oecologia,* 168, 1033-1041.

Afik, O., A. Dag & S. Shafir (2006) The effect of avocado (*Persea americana*) nectar composition on its attractiveness to honey bees (*Apis mellifera*). *Apidologie,* 37, 317-325.

Afik, O., A. Dag & S. Shafir (2008) Honeybee, *Apis mellifera*, round dance is influenced by trace components of floral nectar. *Animal Behaviour,* 75, 371-377.

Benedek, P. (1976) Behaviour of honeybees (*Apis mellifera L.*) in relation to the pollination of onion (*Allium cepa L.*) inflorescences. *Zeitschrift für Angewandte Entomologie* 82, 414-420.

Brewster, J. 2008. *Onions and other vegetable Alliums*. United Kingdom: CAB International.

Brewster, J. L. 1994. Flowering and seed production. In *Onions and Other Vegetable Alliums,* ed. A. J., 122-145. United Kingdom: CAB International.

Briggs, W. H. & I. L. Goldman (2002) Variation in economically and ecologically important traits in onion plant organs during reproductive development. *Plant, Cell and Environment,* 25, 1031-1037.

Brown, J. M., J. A. Wright & R. A. Kohl (1977) Onion-Seed Yield and Quality as Affected by Irrigation Management. *Agronomy Journal,* 69, 369-372.

Butler, C. (1945) The influence of various physical and biological factors of the environment on honey bee activity: an examination of the relationship between activity and nectar concentration and abundance. *The Journal of Experimental Biology,* 21, 5-12.

Chung, I. M., H. K. Song, M. A. Yeo & H. I. Moon (2011) Composition and immunotoxicity activity of major essential oils from stems of *Allium victorialis* L. var. platyphyllum Makino against *Aedes aegypti* L. *Immunopharmacology and Immunotoxicology,* 33, 480-483.

Corzo-Martínez, M., N. Corzo & M. Villamiel (2007) Biological properties of onions and garlic. *Trends in Food Science and Technology,* 18, 609-625.

Currah, L. 1990. Pollination Biology. In *Onion and Allied crops*

Botany, Phisiology, and Genetics, eds. H. Rabinowitch & J. Brewster, 135-149. Washingtong D.C.: CRC Press.

Currah, L. & D. J. Ockendon (1978) Protandry and the sequence of flower opening in the onion (*Allium cepa* L.). *New Phytologist,* 81, 419-428.

Currah, L. & D. J. Ockendon (1984) Pollination activity by blowflies and honeybees on onions in breeders' cages. *Annals of Applied Biology,* 105, 167-176.

Dawid, C., A. Henze, O. Frank, A. Glabasnia, M. Rupp, K. Büning, D. Orlikowski, M. Bader & T. Hofmann (2012) Structural and sensory characterization of key pungent and tingling compounds from black pepper (*Piper nigrum* L.). *Journal of Agricultural and Food Chemistry,* 60, 2884-2895.

Farré-Armengol, G., I. Filella, J. Llusia & J. Peñuelas (2013) Floral volatile organic compounds: Between attraction and deterrence of visitors under global change. *Perspectives in Plant Ecology, Evolution and Systematics,* 15, 56-67.

Fernández, V. A., L. Galetto & J. Astegiano (2009) Influence of flower functionality and pollination system on the pollen size-pistil length relationship. *Organisms diversity & evolution,* 9, 75-82.

Ferreres, F., P. Andrade, M. I. Gil & F. A. Tomás-Barberán (1996) Floral nectar phenolics as biochemical markers for the botanical origin of heather honey. *European Food Research and Technology,* 202, 40-44.

Gegear, R. J., J. S. Manson & J. D. Thomson (2007) Ecological context influences pollinator deterrence by alkaloids in floral nectar. *Ecology Letters,* 10, 375-382.

Gil, M. I., F. Ferreres, A. Ortiz, E. Subra & F. A. Tomás-Barberán (1995) Plant phenolic metabolites and floral origin of rosemary honey. *Journal of Agricultural and Food Chemistry,* 43, 2833-2838.

Hagler, J. R. (1990) Honeybee (*Apis mellifera*) response to simulated onion nectars containing variable sugar and potassium concentrations. *Apidologie,* 21, 115-121.

Hagler, J. R., A. Cohen & G. M. Loper (1990) Production and Composition of Onion Nectar and Honey Bee (Hymenoptera: Apidae) Foraging Activity in Arizona. *Enviromental Entomology,* 19, 327-331.

Havey, M. J. (1999) Seed yield, floral morphology, and lack of male-fertility restoration of male-sterile onion (*Allium cepa*) populations possessing the cytoplasm of *Allium galanthum. Journal of the American Society for Horticultural Science,* 124, 626-629.

Köhler, A., C. W. Pirk & S. W. Nicolson (2012) Honeybees and nectar nicotine: Deterrence and reduced survival versus potential health benefits. *Journal of Insect Physiology,* 58, 286-292.

Kretschmar, J. A. & T. W. Baumann (1999) Caffeine in Citrus flowers. *Phytochemistry,* 52, 19-23.

Kumar, J. & J. Kumar Gupta (1993) Nectar sugar production and honeybee foraging activity in 3 species of onion (*Allium* species). *Apidologie,* 24, 391-396.

Lancaster, J. E. & M. L. Shaw (1989) γ-Glutamyl peptides in the biosynthesis of S-alk(en)yl-l-cysteine sulphoxides (flavour precursors) in *Allium. Phytochemistry,* 28, 455-460.

Long, R. F. & L. Morandin (2011) Low hybrid onion seed yields relate to honey bee visits and insecticide use. *California Agriculture,* 65, 155-158.

Masters, A. R. (1991) Dual role of pyrrolizidine alkaloids in nectar. *Journal of Chemical Ecology,* 17, 195-205.

Mochizuki, E., T. Yamamoto, Y. Komiyama & H. Nakazawa (1998) Identification of *Allium* products using flame photometric detection gas chromatography and distribution patterns of volatile sulfur compounds. *Journal of Agricultural and Food Chemistry,* 46, 5170-5176.

Mogren, L. M., M. E. Olsson & U. E. Gertsson (2006) Quercetin content in field-cured onions (*Allium cepa* L.): Effects of cultivar, lifting time, and nitrogen fertilizer level. *Journal of Agricultural and Food Chemistry,* 54, 6185-6191.

Murali, R. V. N. S. & S. Chandrasekhar (2012) Stereocontrolled synthesis of piperidine alkaloids, (-)-241D and (-)-isosolenopsin. *Tetrahedron Letters,* 53, 3467-3470.

Nicolson, S. W. (2011) Bee food: the chemistry and nutritional value of nectar, pollen and mixtures of the two. *African Zoology* 46, 197-204.

Nicolson, S. W. & R. W. Thornburg. 2007. Nectar Chemistry. In *Nectaries and Nectar*, 215-264. Springer.

Orians, C. M. (2000) The effects of hybridization in plants on secondary chemistry: implications for the ecology and evolution of plant–herbivore interactions. *American Journal of Botany,* 87, 1749-1756.

Parker, F. D. (1982) Efficiency of bees in pollinating onion flowers. *Journal of the Kansas Entomological Society,* 55, 171-176.

Pathak, C. S. (1999) Hybrid Seed Production in Onion. *Journal of New Seeds,* 1, 89-108.

Pérez-Gregorio, M. R., J. Regueiro, C. González-Barreiro, R. Rial-Otero & J. Simal-Gándara (2011) Changes in antioxidant flavonoids during freeze-drying of red onions and subsequent storage. *Food Control,* 22, 1108-1113.

Rabinowitch, H. & J. Brewster. 1990. Pollination biology. In *Onions and Allied crops,* ed. R. J. a. B. J. (eds.). 135-149. USA: CRC Press.

Raguso, R. A. (2004) Why are some floral nectars scented? *Ecology,* 85, 1486-1494.

Saeed, A., M. Taj, K. Abdul, J. Abdul & A. Ghayour (2003) Effect of pollinators and insecticides on seed setting of onion (*Allium cepa* L.). *Journal of Biological Sciences,* 3, 305-308.

Sajjad, A., S. Saeed & A. Masood (2008) Pollinator community of onion (*Allium cepa* L.) and its role in crop reproductive success. *Pakistan Journal of Zoology,* 40, 451-456.

Scott, I. M., N. Gagnon, L. Lesage, B. J. R. Philogène & J. T. Arnason (2005) Efficacy of botanical insecticides from Piper species (*Piperaceae*) extracts for control of European chafer (Coleoptera: *Scarabaeidae*). *Journal of Economic Entomology,* 98, 845-855.

Silva, E. M. & B. B. Dean (2000) Effect of nectar composition and nectar concentration on honey bee (Hymenoptera: Apidae) visitations to hybrid onion flowers. *Journal of Economic Entomology,* 93, 1216-1221.

Silva, E. M., B. B. Dean & L. Hiller (2004) Patterns of floral nectar production of onion (*Allium cepa* L.) and the effects of environmental conditions. *Journal of the American Society for Horticultural Science,* 129, 299-302.

Silva, E. M., B. B. Dean & L. K. Hiller (2003) Honey Bee (Hymenoptera: *Apidae*) foraging in response to preconditioning with onion flower scent compounds. *Journal of Economic Entomology,* 96, 1510-1513.

Soto, V., I. Maldonado, V. Jofre, C. Galmarini & M. F. Silva. (unpublished). Are nectar and floral volatile organic compounds related to pollination and seed production of onion F1 hybrids?

Soto, V. C., I. B. Maldonado, R. A. Gil, I. E. Peralta, M. F. Silva & C. R. Galmarini (2013) Nectar and flower traits of different onion male sterile lines related to pollination efficiency and seed yield of F1 hybrids. *Journal of Economic Entomology,* 106, 1386-1394.

Storsberg, J., H. Schulz, M. Keusgen, F. Tannous, K. J. Dehmer & E. R. J. Keller (2004) Chemical characterization of interspecific hybrids between *Allium cepa* L. and *Allium kermesinum* Rchb. *Journal of Agricultural and Food Chemistry,* 52, 5499-5505.

Torres, C. & L. Galetto (1998) Patterns and implications of floral nectar secretion, chemical composition, removal effects and standing crop in *Mandevilla pentlandiana* (Apocynaceae). *Botanical Journal of the Linnean Society,* 127, 207-223.

Truchado, P., F. Ferreres, L. Bortolotti, A. G. Sabatini & F. A. Tomás-Barberán (2008) Nectar flavonol rhamnosides are floral markers of acacia (*Robinia pseudacacia*) honey. *Journal of Agricultural and Food Chemistry,* 56, 8815-8824.

Waller, G. (1974) Evaluating the foraging behavior of honey bees as pollinators of hybrid onions. *Proc Third Int Pollination Symp Prague,* 175-180.

Waller, G., E. Carpenter & O. Ziehl (1972) Potassium in onion nectar and its probable effects on attractiveness of onion flowers to honey bees. *Journal of American Society of Horticultural Science,* 97, 535-539.

Waller, G., N. Waters, E. Erickson & J. H. Martin (1976) The use of Potassium to identify onion nectar-collectin honey bees. *Environmental Entomology,* 5, 780-782.

Wilkaniec, Z., K. Giejdasz & G. Prószyñski (2004) Effect of pollination seeds under isolation by the red mason bee (*Osmia rufa* L.) (*Apoidea, Megachilidae*) on the setting and quality of obtained seeds. *Journal of Apicultural Science,* 48, 35-41.

Xue, R. D., D. R. Barnard & A. Ali (2001) Laboratory and field evaluation of insect repellents as larvicides against the mosquitoes *Aedes albopictus* and *Anopheles albimanus. Medical and Veterinary Entomology,* 15, 374-380.

In: Nectar
Editor: Robin L. Peck
ISBN: 978-1-63463-650-6

Chapter 3

EXTRAFLORAL NECTAR PRODUCTION AND PLANT DEFENSE STRATEGIES

Akira Yamawo[1] and Makoto Tokuda[2]
[1]Department of Biology, Faculty of Science, Kyushu University, Higashi Ward, Fukuoka, Japan
[2]Faculty of Agriculture, Saga University, Saga, Japan

ABSTRACT

In plant-ant mutualism, plants often provide foods, such as extrafloral nectar and food bodies, to ants. In return, the visiting ants protect the plants against herbivores. However, these plants also possess various direct defenses, including physical defense via trichomes and tough leaves, chemical defense using secondary metabolites. Why have plants evolved defensive mutualisms with ants even when possessing these direct defenses? To answer this question, we must clarify the costs and benefits of each defense trait, as well as its role in the overall defense strategy of the plant. Previous studies, however, have mostly dealt with the costs and benefits of particular defenses. Here, we review recent advances in the study of plant defense strategies involving the indirect defense provided by extrafloral nectar production and other defenses. First, we explain the benefits of each defense trait, with special reference to its efficacy against herbivores. Second, we review the costs of each defense trait. Next, we focus on differences in the use of each defense trait in response to habitat conditions and the value of plant tissue.

Finally, we highlight the major gaps in our current knowledge and outline future research directions.

INTRODUCTION

Because autotrophic green plants produce the bulk of the energy that supports other organisms, it should not be surprising that plant defenses against herbivores comprise a prominent set of adaptive measures in the history of life [1]. To protect their own resources, plants have evolved direct defense traits against herbivory [2, 3].

Direct defenses include physical structures, such as trichomes and tough leaves, and chemical substances, such as alkaloids and tannins. In addition, many plants have evolved indirect biotic defenses, i.e., extrafloral nectaries [EFNs] and food bodies [lipid-rich particles; FBs] that attract natural enemies of herbivores, such as ants and parasitic wasps [4, 5].

EFNs have been identified in approximately 4,000 species worldwide, with examples in 745 genera from 108 families of flowering plants [6].

A previous phylogenetic study has suggested multiple evolutionary origins of EFNs in at least 457 independent lineages [6]. EFNs have been described in virtually all aboveground plant tissues and organs, including leaves, petioles, stipules, cotyledons, fruits, and stems.

Some EFN-bearing plants also possess direct defense traits [7, 8, 9, 10, 11]. For example, *Paulownia tomentosa*, *Conocarpus erectus*, and *Mallotus japonicus* use trichomes as effective physical defenses [9, 11, 12]. *Cnidocampa flavescens* and *Gossypieae* plants possess chemical defensive substances against herbivores [8, 13, 14].

To ascertain why these plants have evolved indirect defense traits despite their possession of direct defenses, comprehensive assessments of the respective costs, benefits, and uses of direct and indirect defenses are very important. However, previous studies have mostly examined the costs and benefits of particular indirect defenses in isolation [1, 15, 16, 17].

Here, we review recent advances in the study of plant defense strategies involving the indirect defense provided by EFN production. First, we explain the benefits of each defense trait, with special reference to its efficacy against herbivores. Second, we summarize the costs of each defense trait. Third, we focus on differences in the use of each defense trait in response to habitat conditions and the value of plant tissue. Finally, we highlight the major gaps in our current knowledge and discuss future perspectives.

Efficacy of Direct and Indirect Defenses

Plants generally possess direct defense traits, including both physical and chemical defenses, for avoiding herbivory. Numerous studies have demonstrated the function of direct defense traits against herbivores [reviewed in 2, 3, 18]. For example, trichomes on plant surfaces or tough cell walls physically protect plant tissues from herbivores [18]. Chemical substances are also known to defend plant tissues against herbivores [reviewed in 2, 3, 18]; these compounds are classified into quantitative substances such as tannins and phenolic compounds and qualitative ones such as alkaloids [2, 19, 20]. Chemical defenses are effective against numerous herbivores, but their efficacy depends on herbivore species or type. In general, the efficacy of chemical defenses is lower against specialist herbivores than against generalist herbivores, because specialists are often physiologically adapted to the chemicals synthesized by their host plants [2, 20, 21]. Therefore, many previous studies have accepted the notion that specialists are immune to host plant defenses. However, this view may not be applicable for some systems. For example, negative effects of host chemical defenses have been detected on the performance of specialist herbivores in systems such as parsnip webworms [*Depressaria pastinacella*] intaking host furanocoumarins, buckeye caterpillars [*Junonia coenia*] ingesting iridoid glycosides, monarch caterpillars [*Danaus plexippus*] feeding on cardenolide-containing sandhill milkweed [*Asclepias humistrata*], cabbage white caterpillars [*Pieris rapae*] consuming isothiocyanates, and tobacco hornworms [*Manduca sexta*] fed on artificial diets containing nicotine [reviewed in 22]. Thus, higher levels of chemical defenses are effective even against specialists [22].

Many studies have revealed that EFNs function in a range of plant species as indirect defenses by inducing ants to combat herbivores. A meta-analysis by Chamberlain and Holland [23] showed that the effects of EFNs and ants on plants are generally positive, as ants reduce herbivory and increase plant performance. However, a defensive function has not always been demonstrated for EFNs, and some studies have failed to detect a deterring effect of visiting ants on herbivores [e.g., 24, 25, 26, 27, 28]. The number of ants on a plant and ant attendance rate often vary with the time of day, season, and habitat [29, 30, 31, 32, 33, 34]. Thus, EFN-bearing plants sometimes fail to attract ants. For example, in *Vicia sativa* var. *angustifolia* and *M. japonicus*, 35.6% and 15% of individuals, respectively, lacked ants [34, 35]. In addition, the species composition of ants, which is an important factor for the efficacy of indirect defenses, also varies due to habitat conditions such as light, soil moisture, and disturbance [34]. Mody and Linsenmair [28] and Miller [36] have demonstrated that herbivory levels significantly differ among plants dominated by different ant species.

Therefore, unlike that of physical or chemical defenses, the efficiency of indirect defenses via EFNs is rather unstable and depends on the abundance and species of responsible ants [31, 34, 37].

To date, few studies have focused on differences in the efficacies of direct and indirect defenses. When EFN-bearing plant species are grown with other species that use direct defenses, the former are generally consumed to a greater degree by herbivores [35]. Similarly, plants expressing indirect defense phenotypes are more often consumed by herbivores than those expressing direct defenses [12, 34]. In addition, leaf herbivory is greater in leaves that express indirect defenses than in those expressing physical and chemical defenses within the same plant [38, 39].

Direct defenses seem to be generally more effective against herbivores than are indirect defenses via EFNs. However, indirect defenses by ants may be superior in some aspects to other direct and indirect defenses. For example, ants might serve as effective defenders against a wide range of insect herbivores, such as specialists that have adapted to direct defenses [40, 41]. Herbivores likely have more difficulty in overcoming indirect defenses compared with direct defenses [42]. Further studies must be conducted to reveal differences in the efficacy of direct and indirect defenses.

COSTS OF DEFENSE TRAITS

Physiological and Allocation Costs

Because the costs of defenses provide the basis for fundamental ecological and evolutionary hypotheses regarding plant defense strategies [29, 43, 44, 45, 46, 47], many studies have attempted to estimate these costs, some more successfully than others [reviewed in 48, 49, 50]. Allocation costs occur when a limited resource is used for defense at the expense of primary metabolism, growth, or reproduction [51]. Many earlier works have shown that the expression of physical and chemical defense traits impedes plant growth and reproduction [50, 52, 53, 54, 55]. For example, Feng et al. [54] have reported that the resource allocation of nitrogen to cell walls decreases the allocation to photosynthesis [i.e., plant growth]. In tomato *Solanum lycopersicum* [Solanaceae], the induction of chemical defense substances such as polyphenol oxidase and peroxidase by jasmonic acid decreased seed production by 33% [53]. In contrast, the expression of EFNs and production of extrafloral nectar seem to have few allocation costs [16, 56, 57]. Rutter and Rausher [57] have reported that extrafloral nectar production in *Chamaecrista fasciculata* did not reduce plant growth or seed production. In addition, ant attendance for this species had no influence on plant growth [16].

Such differences between the costs of direct and indirect defenses have also been found in several studies examining quantitative costs. Cellulose and hemicellulose, which stiffen leaves, account for 80%–90% of leaf dry-weight [58]. In *Thea* = *Camellia sinensis*, the cost of secondary compound production accounts for 30% of leaf dry weight [59]. In the myrmecophyte *Macaranga tanarius* [Euphorbiaceae], the phenolic content represents approximately 20% of the leaf biomass [60]. In contrast, the cost of EFN production is as low as approximately 1% of the total energy invested in leaves in *Ochroma pyramidale* [15] or approximately 0.07% of total plant biomass in *V. sativa* [35]. Thus, the allocation costs of EFNs and extrafloral nectar are likely negligible, in contrast to the relatively high allocation costs of direct defense traits. Differences in the costs of direct (physical/chemical) and indirect defenses must reflect differences in the materials used for these respective defense traits. Physical and chemical defense traits are often constructed using cellulose/hemicellulose and chemical metabolites, respectively.

Trichomes, for example, consist fundamentally of cellulose [61], of which the synthesis costs are comparatively high [62] compared to those of the primary sugars in extrafloral nectar [35, 38]. However, no studies have rigorously quantified the direct metabolic costs of extrafloral nectar [63].

Earlier studies have examined the costs and benefits of particular defenses in various plants [1, 15, 16, 17] but are limited by the lack of comparisons within plant species. The costs of defense traits may depend on resource arability, resource capacity, phylogenetic limitations, and metabolism in each plant species [64].

Thus, it is critically important to estimate costs within a species to understand the differences in the costs of direct and indirect defenses.

Ecological Costs

Ecological costs result from the negative effects of defenses on any interactions between a plant and other organisms that would affect its fitness under natural conditions, e.g., the deterrence of mutualists, increased parasite load or enemy impacts, reduced tolerance to enemies, and reduced competitive ability in the environment [3, 50].

A number of studies have sought empirical evidence for the ecological costs of direct defense traits [reviewed in 50]. For example, trade-offs among direct defense traits have been reported in many plants [8, 65, 66, 67, 68, 69, 70]. A phylogenetically controlled analysis of 31 cotton [*Gossypium*] species found a trade-off between the direct defense traits of leaf glands and trichomes [8]. Other studies have shown that the expression of direct defense traits reduces the indirect defense of volatile emissions [71, 72] or tolerance to leaf damage [73, 74, 75]. A recent study using *Phaseolus lunatus* demonstrated a trade-off between direct defense by the release of hydrogen cyanide [HCN] and indirect defense by the emission of volatiles [72]. Trade-offs between direct defense and competitive ability have also been observed. For example, the induction of chemical defenses in *Nicotiana sylvestries* reduces its competitive ability for nitrogen, resulting in an approximately 50% decrease in seed production [76]. Ballhorn et al. [77] have also demonstrated that a highly defended lima bean genotype produced significantly fewer seeds than a less-defended genotype under competitive conditions.

Research has also shown that a certain trait can have multiple ecological functions that pose conflicting problems for plants. For example, the production of chemical defense substances deters generalist herbivores but also attracts specialist herbivores [78]. Additionally, plants that are resistant to herbivores due to the accumulation of toxic secondary compounds for defense, with higher concentrations of these compounds in their fruit, floral tissues, and nectar than in those of susceptible plants, may be less attractive to pollinator or disperser mutualists [63, 79].

Indirect defenses using EFNs also present several ecological costs. The occupation of EFNs by nectar-robbing insects could prevent mutualistic ants from exhibiting their defensive effects [4, 80, 81, 82]. In addition, visitations by aggressive ants reduce the frequency of plant-pollinator interactions, and hence plant reproduction [32, 83, 84, 85].

Although trade-offs between direct defenses and indirect defenses by ants have been observed in obligate ant-plant mutualism [86; 87, 88, 89; 90, 91 but see 7, 82, 92], they have never been confirmed in facultative ant-plant mutualism via EFNs [8]. A phylogenetically controlled analysis of 31 *Gossypium* species did not find a trade-off between EFNs and toxic glands/ trichomes [8]. In addition, based on a meta-analysis of 25 studies, Koricheva and Romero [70] found a significant negative correlation between plant allocation to direct and ant-mediated defenses.

However, these trade-offs were significant only in plants that offered more costly rewards to ants, such as food bodies and/or domatia, and not in those that offered EFNs. These results suggest that plants have evolved EFNs in conjunction with direct defense traits [7, 8, 9, 10, 39] due to the low production costs of EFNs.

How to Use Each Defense ?

Leaf Development

Plants can alter the intensities of antiherbivore defenses depending on the developmental stages of their organs. That plants exhibit optimal resource allocation to each defense has been proposed by McKey [93, 94] and Rhoades [44] as the "optimal defense theory." The optimal defense strategy is determined by factors such as the value of the defended organs, physiological cost of defense traits, and herbivory pressures. Because these factors undergo changes during the growth and aging of leaves [95, 96, 97], leaf age is closely associated with the optimal defense strategy [97, 98, 99, 100].

To date, many studies have shown variations in direct defense traits, such as trichomes, secondary compounds, and leaf toughness, with leaf age [9, 95 97]. Many plant species use trichomes mainly on young leaves to diminish herbivore activity, as growing young leaves have high assimilative abilities and high nutrient concentrations [97, 98, 100, 101] but limited physical toughness. New leaves are inevitably soft because they must allow cell expansion and proliferation.

Several studies of fast-growing plants have demonstrated that chemical substances are also accumulated in young plants and decrease with leaf age [e.g., 93, 102] as they are reallocated from mature leaves to new ones [102].

Indirect defense traits, including EFNs, usually function in young to middle-aged leaves [15, 89, 101, 103]. Extrafloral nectar production is most intensive in these leaves, although they have not yet reached their maximum rates of photosynthetic assimilation [89]. These results indicate that plants alter their expression of extrafloral nectar in response to ecological requirements through the transportation of resources. In fact, lima bean and *Ricinus communis* secrete extrafloral nectar in younger leaves due to the transportation of newly assimilated carbohydrates from mature leaves [101]. This pattern matches the spatial and temporal distribution of defense investment predicted by the optimal defense theory [44, 93].

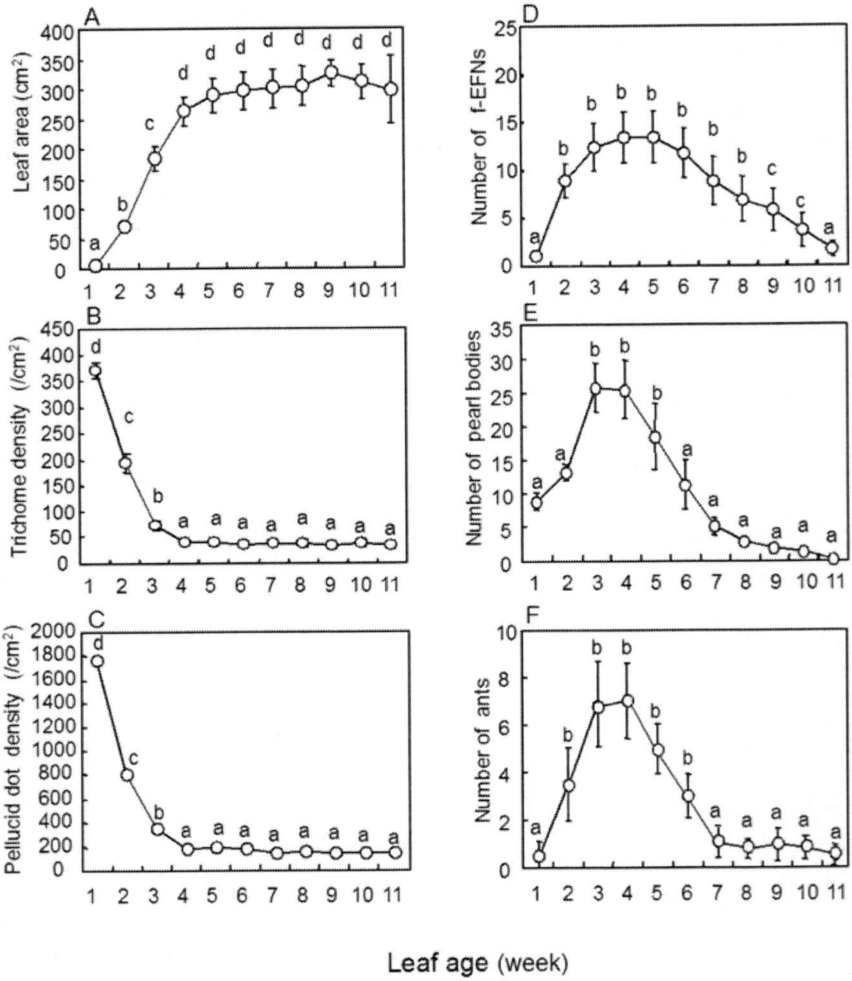

Figure 1. Leaf area, the density or number of defense traits on leaves from 1 to 11 weeks, and the number of ants on leaves of each age. [A] leaf area; [B] trichome density; [C] pellucid dot density; [D] number of functional extrafloral nectaries; [E] number of pearl bodies; [F] number of ants. Bars represent standard error [SE]. Different letters denote significant differences [Steel-Dwass, $P < 0.05$] [39].

In recent years, several studies have focused on the use of multiple defense traits in plants and estimated the expression of both direct and indirect defenses in different parts. For example, Yamawo et al. [39] reported a shift in defense tactics from direct (using trichomes and pellucid dot chemicals) to indirect (using EFNs and FBs) in relation to the development and aging of leaves in the pioneer plant *Mallotus japonicus* [Figure 1]. Similar phenomena have also been observed in other fast-growing pioneer trees such as *Macaranga tanarius* [89, 104] and *Paulownia tomentosa* [9].

The shift from direct to indirect defenses agrees with the optimal defense theory. In general, the value of young leaves is higher than that of mature leaves [97, 100, 101]. This situation is also applicable for *M. japonicus*, because the removal of young leaves decreases plant biomass to a greater degree than does the removal of mature ones [unpublished data]. Young leaves must also be more vulnerable even to slight herbivory because of their smaller areas. Therefore, repelling herbivores from young leaves is a matter of priority for *M. japonicus*. As mentioned above, the efficacy of indirect defenses varies according to habitat conditions because of its dependence on the abundance of ants nearby [31, 34, 39]; the effect of indirect defenses is thus unstable. In contrast, the efficacy of direct defenses is more stable because their production directly affects herbivores. For these reasons, plants are expected to express direct defense traits to a greater degree in valuable young or small leaves. In addition, several authors have suggested that the physiological cost of producing indirect defense traits is lower than that of producing direct defense traits [15, 35]. Therefore, the allocation of limited resources to direct defenses for both young and mature leaves may be too costly for some plants. This situation is particularly important for pioneer plants, which are exposed to heavy competition for light and are thus required to grow rapidly. For such plants, saving resources by adopting low-cost indirect defenses results in greater resource allocation to rapid growth. Therefore, the protection of less valuable mature leaves by indirect defenses helps plants achieve higher growth rates and is thus adaptive.

Such shifts in defense tactics related to leaf age occur differently even among closely related plant species depending on their life histories, which also affect optimal defense strategies [105, 106]. For example, both leaf age and leaf longevity impact the adaptive defense tactics of plants, with long-living leaves protected more intensively than short-living leaves [98, 106, 107, 108]. Several studies have predicted that, although fast-growing plants should defend young leaves more efficiently than they defend mature leaves [93, 102], slowly growing plants must defend mature leaves just as effectively as they do young leaves [98, 108] because of their greater leaf longevity.

Relatively few studies to date have surveyed the expression of direct and indirect defenses along with changes in leaf age in nonpioneer plants with low growth rates.

The slow-growing and shade-tolerant plant *Mallotus philippensis* employs direct chemical defenses in leaves of all ages, accompanied by a shift from indirect defense using EFN in young leaves to direct physical defense using leaf toughness in middle-aged leaves [109]. Another slow-growing species, *Copaifera langsdorffii*, also uses chemical defenses (tannins) in both young and mature leaves but shifts from indirect defense by EFNs in young leaves to direct defense by leaf toughness in mature leaves [110]. The continuous investment in the chemical defenses of leaves by slow-growing plants also supports the optimal defense theory [98, 108]. These studies strongly suggest that, in relation to their life histories, EFN-bearing plants adopt different defense strategies using multiple defense traits for different leaf ages and longevities.

Habitat Conditions

Plants can also change the intensities of their antiherbivore defenses depending on abiotic habitat conditions such as light, soil moisture, and nutrients. Many hypotheses have been proposed to understand plant defense strategies in relation to abiotic conditions. The "growth-differentiation balance hypothesis" [GDB] [29] and "resource availability hypothesis" [RA] [45] predict that resource availability will exert an effect on the expression of defense traits [29, 45]. Fast-growing plants under high nutrient conditions are predicted have low levels of defense, because the production of new leaves is supported by the exportation of photosynthates from source leaves, leaving little energy for the synthesis of defense traits. The results of several studies have supported the GDB and RA hypotheses. For example, plants generally express direct defense traits such as chemical defensive substances, dense trichomes, and leaf toughness to a greater degree under nutrient-poor conditions [e.g., 111, 112], and soil fertilization decreases leaf phenolic contents [111]. The "carbon-nutrient balance" [CNB] hypothesis assumes that plants will invest preferentially in carbon-based defenses when soil nutrients rather than photosynthesis limit their growth. However, in situations where photosynthesis is restricted by factors other than soil nutrients, these surplus nutrients will be allocated to defense traits [113]. Numerous previous studies have examined the CNB hypothesis, but most of them have obtained equivocal results [114], with only several supporting hypothesis [3, 114]. For example, in many plant species soil fertilization decreases the concentration of non-nitrogen–containing secondary metabolites but increases that of nitrogenous secondary metabolites [reviewed in 114].

Several studies have investigated the effects of abiotic conditions on indirect defenses such as EFNs and food bodies. Soil fertilization, or increased soil nutrient contents, increases the production of EFNs and food bodies [115, 116, 117]. Folgarait and Davidson [115] and Heil et al. [116] have reported the positive effects of nutrient supply on the production of food bodies in *Cecropia* species and *Macaranga triloba*. *Acacia constricta* grown by an ant nest had more EFNs per leaf than did plants without nearby ant nests, as these nests provided soil nutrients for the plants [117].

Light conditions also affect the expression of indirect defense traits. In general, carbon-based indirect defense traits, such as EFNs and the Müllerian bodies of *Cecropia* plants, are expressed to a greater degree under high-light conditions, resulting in stronger ant attraction [118, 119, 120], but food bodies that contain large amounts of nitrogen are expressed to a greater degree under low-light conditions. These results support the CNB hypothesis. In another EFN-bearing plant, *Inga vera*, the effects of ants on plant growth have been detected in high-light conditions [121].

Conversely, Radhika et al. [122] have demonstrated that complex interactions occur between light conditions and extrafloral nectar production in lima bean. Control plants without herbivory secreted a large amount of extrafloral nectar under low -light conditions, but treatment plants in which antiherbivory defenses were activated through the addition of jasmonic acid [JA] secreted extrafloral nectar under high-light conditions.

Although most previous studies have focused on the plasticity of either direct or indirect defenses in relation to abiotic conditions, studies of young *Mallotus japonicus* plants examined both types. Yamawo and Hada [119] reported that an increase in EFN size and decrease in food bodies occurred under high-light conditions in young plants of these species. Conversely, Yamawo et al. [11] and Yamawo et al. [123] demonstrated that, under high soil nutrient and high moisture conditions, young *M. japonicus* increased EFNs and extrafloral nectar volume while decreasing the densities of trichomes and pellucid dots. Moreover, Yamawo et al. [34] showed that these plants express different defense strategies in response to habitat conditions in the field. At open sites with conditions of high light and low soil nutrients/moisture, trichomes and pellucid dots were expressed to greater degree than at treefall gaps, which had conditions of low light and high soil nutrients/moisture. In fact, leaves collected from the open sites were consumed less by the generalist herbivore, *Spodoptera litura*, than were those from other habitats. This result indicates that direct defenses function to a greater degree at open sites.

Although the contents of the pellucid dots in *M. japonicus* have not been identified, those of an allied species, *Macaranga tanarius*, are known to contain the prenylated flavanone nymphaeol-C [104].

Trichomes and pellucid dots are carbon-based defense traits thought to be closely related to photosynthetic activity.

At open sites, *M. japonicus* favors carbon-based physical and chemical defenses, such as trichomes and pellucid dots, because the high light availability [111]. These results support the CNB hypothesis [113].

The low expression of direct defense traits at forest edges and treefall gaps may be attributed to their high production costs. At these habitats, *M. japonicus* must compete more intensively for light than at open sites. Under restricted light conditions, the expression of physical and chemical defense traits is costly for non-shade–tolerant plants such as herbs and pioneer tree species. In fact, Kitajima et al. [124] have suggested that carbon allocation to cell walls [i.e., leaf toughness] is costly, particularly in non-shade–tolerant plants. Many reports have shown that the expression of trichomes, leaf toughness, and chemical substances can reduce plant growth [e.g., 50, 53] and that plants expressing high chemical defenses often possess low competitive ability [77]. Therefore, at forest edges and treefall gaps, non-shade–tolerant plants are expected to decrease their expression of direct defense traits and allocate more resources to their own growth [125, 126]. The adoption of low-cost indirect defenses contributes not only to defensive functions but also to the increase of resource allocation to growth.

Conclusion

Many previous studies have demonstrated the defensive functions of EFNs in various plant species, but the differences in costs and benefits have not yet been sufficiently compared between EFNs and other defenses. Several studies have strongly suggested that the cost and efficacy of indirect defense by EFNs are lower and more unstable, respectively, than those of direct defenses [39]. However, these studies have focused only on particular plant species, such as *M. japonicus* and *V. sativa*. Further studies are necessary to confirm the universality of these results by examining various systems. In this paper, we have reviewed the use of EFNs and other defenses in relation to the value of plant tissue and habitat conditions.

This work provides a comprehensive overview of both plant defense strategies and plant-ant mutualisms. Future studies on the use of EFNs and other defense traits in response to other biotic factors, such as communities of herbivores, mutualists, and competitors, must be conducted to understand the ecology and evolution of EFNs in plants.

REFERENCES

[1] A. Agrawal, *Funct. Ecol.* 25, 420–432 (2011).

[2] H. F. Howe and L. C. Westley, *Ecological Relationships of Plants and Animals*. Oxford University Press, New York (1988).

[3] D. R. Walters, *Plant Defense: Warding off Attack by Pathogens, Herbivores and Parasitic Plants*. Blackwell Publishing, Iowa, IA. (2011).

[4] S. Koptur, Extrafloral nectary-mediated interactions between insects and plants. *Insect–Plant Interactions*, Vol. 4. (ed. E. A. Bernays), pp. 81–129. CRC Press, Boca Raton. (1992).

[5] V. Rico-Gray and P. S. Oliveira, *The ecology and evolution of ant–plant interactions*. The University of Chicago Press, Chicago (2007).

[6] M. G. Weber and K. H. Keeler, *Ann. Bot.* 111, 1251–1261 (2013).

[7] J. L. Steward and K. H. Keeler, *Oikos* 53, 79–86 (1988).

[8] J. A. Rudgers, S. Y. Strauss and J. F. Wendel, *Am. J. Bot.* 91, 871–880 (2004).

[9] S. Kobayashi, T. Asai, Y. Fujimoto, and S. Kohshima, *Ann. Bot.* 101, 1035–1047 (2008).

[10] Young, D. Wagner, P. Doak and T. Clausen, *Ame. J. Bot.* 97, 601–610 (2010).

[11] Yamawo, N. Katayama, N. Suzuki and Y. Hada, *Plant Ecol.* 213, 127–132 (2012b).

[12] J. Piovia-Scott, *J. Ecol.* 99, 327–334 (2011).

[13] W. Huang, E. Siemann, G. S. Wheeler, J. Zou, J. Carrillo, and J. Ding, *J. Ecol.* 98, 1157–1167 (2010).

[14] X. Wang, L. P. Yomano, J. Y. Lee, S. W. York, H. Zheng, M. T. Mullinnix, K. T. Shanmugam, and L. O. Ingram, *Proc. Natl. Acad. Sci.* 110, 4021–4026 (2013).

[15] J. O'Dowd, *Oecologia* 43, 233–248 (1979).

[16] M. Barton, *Ecology* 67, 495–504 (1986).

[17] M. Heil, B. Fiala, K. E. Linsenmair, G. Zotz, P. Menke, and U. Maschwitz, *J. Ecol.* 85, 847–861 (1997).
[18] A. Mithöfer and W. Boland, *Annu. Rev. Plant Biol.* 63, 431–450 (2012).
[19] G. A. Rosenthal and D. H. Janzen, *Herbivores. Their Interaction with Secondary Plant Metabolites.* Academic Press, New York (1979).
[20] J. A. Gatehouse, *New Phytol.* 156, 145–169 (2002).
[21] E. van der Meijden, In: *Proceedings of the 9th International Symposium on Insect-Plant Relationships* (pp. 307-310). Springer Netherlands (1996).
[22] J. G. Ali and A. A. Agrawal, *Trends Plant Sci.* 17, 293–302 (2012).
[23] S. A. Chamberlain and J. N. Holland, *Ecology* 90, 2384–2392 (2009).
[24] J. O'Dowd and E. A. Catchpole, *Oecologia* 59, 191–200 (1983).
[25] D. A. Mackay and M. A. Whalen, *J. Entomol.* 37, 335–339 (1998).
[26] S. Kawano, H. Azuma, M. Ito, and K. Suzuki, *Plant Species Biol.* 14, 167–178 (1999).
[27] L. Freitas, L. Galetto, G. Bernardello, and A. A. Paoli, *Flora (Jena)*, 195, 398–402 (2000).
[28] K. Mody and K. L. Linsenmair, *Ecol. Ent.* 29, 217–225 (2004).
[29] D. A. Herms and W. J. Mattson, *Quarterly Rev. Biol.* 67, 283–335 (1992).
[30] J. L. Bronstein, *Biotropica* 30, 150–161 (1998).
[31] B. D. Giusto, M. C. Anstett, E. Dounias, and D. B. McKey, *Oecologia* 129, 367–375 (2001).
[32] J. H. Ness, *Oikos* 113, 506–514 (2006).
[33] J. A. Rudgers, A. M. Savage and M. A. Rúa, *Oecologia* 163, 985–996 (2010).
[34] A. Yamawo, J. Tagawa, Y. Hada, and N. Suzuki. *J. Ecol.* 102, 238–247 (2014).
[35] N. Katayama and N. Suzuki, *Plant Ecol.* 212, 743–752 (2011).
[36] T. M. Miller, *Oikos* 116, 500–512 (2007).
[37] J. Ness, *Oecologia* 134, 210–218 (2003).
[38] M. Heil, T. Koch, A. Hilpert, B. Fiala, W. Boland, and K. E. Linsenmair, *Proc. Natl. Acad. Sci.* 98, 1083–1088 (2001).
[39] A. Yamawo, N. Suzuki, J. Tagawa, and Y. Hada, *J. Ecol.* 100, 802–809 (2012).
[40] K. H. Keeler, *Ame. J. Bot.* 64, 1182–1188 (1977).
[41] R. M. Beach, J. W. Todd and S. H. Baker, *J. Entomol. Sci.* 20, 233–236 (1985).

[42] J. Kahl, D. H. Siemens, R. J. Aerts, R. Gäbler, F. Kühnemann, C. A. Preston, and I. T. Baldwin, *Planta* 210, 336–342 (2000).
[43] P. Feeny, *Biochemical interaction between plants and insects*. Springer US, 1–40 (1976).
[44] D. F. Rhoades, *Evolution of plant chemical defense against herbivores. Herbivores: Their Interactions with Secondary Plant Metabolites*. pp. 3–54. Academic Press, New York, Orlando, FL. (1979).
[45] P. D. Coley, J. P. Bryant and F. S. III. Chapin, *Science* 230, 895–899 (1985).
[46] J. Gershenzon, *J. Chem. Ecol.* 20, 1281–1328 (1994).
[47] T. E. Ohnmeiss and I. T. Baldwin, *Ecology* 81, 1765–1783 (2000).
[48] J. Bergelson and C. B. Purrington, *Ame. Nat.* 148, 536–558 (1996).
[49] J. Koricheva, *Ecology* 83, 176–190 (2002).
[50] S. Y. Strauss, J. A. Rudgers, J. A. Lau, and R. E. Irwin, *Trends Ecol. Evol.* 17, 278–285 (2002).
[51] M. Heil and I. Baldwin, *Trends Plant Sci.* 7, 61–67 (2002).
[52] A. Agrawal, *Ecology* 81, 1804–1813 (2000).
[53] M. Redman, D. F. Jr. Cipollini and J. C. Schultz, *Oecologia* 126, 380–385 (2001).
[54] Y. L. Feng, Y. B. Lei, R. F. Wang, R. M. Callaway, A. Valiente-Banuet, Li. Y. P. Inderjit, and Y. L. Zheng, *Proc. Natl. Acad. Sci.* 106, 1853–1856 (2009).
[55] M. Orians, C. G. Hochwender, R. S. Fritz, and T. Snäll, *Oecologia* 163, 283–290 (2010).
[56] K. Adjei-Maafo, L. T. Wilson, N. J. Thomson, and P. R. B. Blood, *Environ. Entomol.* 12, 353–358 (1983).
[57] M. T. Rutter and M. D. Rausher, *Evolution* 58, 2657–2668 (2004).
[58] P. J. van Soest, *Nutritional Ecology of the Ruminant*. O and B Books, Corvallis, Oregon. (1982).
[59] M. L. Vickery and B. Vickery, *Secondary Plant Metabolism*. Macmillan, London (1981).
[60] T. Y. Lim, Y. Y. Lim and C. M. Yule, *Food Chem.* 114, 594–599 (2009).
[61] L. Betancur, B. Singh, R. A. Rapp, J. F. Wendel, M. D. Marks, A. W. Roberts, and C. H. Haigler, *J. Integrative Plant Biol.* 52, 205–220 (2010).
[62] P. Scott, *Physiology and Behaviour of Plants*. Chichester: John Wiley and Sons, Ltd. (2008).

[63] A. Kessler, R. Halitschke and K. Poveda, *Ecology* 92, 1769–1780 (2011).
[64] D. Houle, *Evolution* 45, 630–648 (1991).
[65] R. Zangerl and M. R. Berenbaum, *Ecology* 71, 1933–1940 (1990).
[66] K. Brody and R. Karban, *Oikos* 65, 301–306 (1992).
[67] R. Mauricio, M. D. Rausher and D. S. Burdick, *Ecology* 78, 1301–1311 (1997).
[68] J. S. Thaler, R. Karban, D. E. Ullman, K. Boege, and R. M. Bostock, *Oecologia* 131, 227–235 (2002).
[69] D. Ward and T. P. Young, *J. Chem. Ecol.* 28, 921–937 (2002).
[70] J. Koricheva and G. Q. Romero, *Biol. Let.* 8, 628–630 (2012).
[71] A. Agrawal, A. Janssen, J. Bruin, M. A. Posthumus, and M. W. Sabelis, *Ecol. Let.* 5, 377–385 (2002).
[72] J. Ballhorn, S. Kautz, U. Lion, and M. Heil, *J. Ecol.* 96, 971–980 (2008).
[73] W. L. Fineblum and M. D. Rausher, *Nature* 377, 517–520 (1995).
[74] K. A. Stowe, R. J. Marquis, C. G. Hochwender, and E. L. Simms, *Annu. Rev. Ecol. Evol. and Syst.* 31, 565–595 (2000).
[75] P. Tiffin and M. D. Rausher, *Ame. Nat.* 154, 700–716 (1999).
[76] T. Baldwin and Iii, W. Hamilton, *J. Chem. Ecol.* 26, 915–952 (2000).
[77] D. J. Ballhorn, A. L. Godschalx, S. M. Smart, S. Kautz, and M. Schädler, *Oecologia* 1–14 (2014).
[78] A. Agrawal and M. F. Sherriffs, *Ann. Entomol. Soc. Ame.* 94, 71–75 (2001).
[79] L. S. Adler, *Oikos* 91, 409–420 (2000).
[80] J. Beattie, *The evolutionary ecology of ant-plant mutualisms*. Cambridge University Press (1985).
[81] M. Heil, A. Hilpert, R. Krüger, and K. E. Linsenmair, *J. Tropical Ecol.* 20, 201–208 (2004).
[82] M. Heil, T. Delsinne, A. Hilpert, S. Scürkens, C. Andary, and K. E. Linsenmair, *Oikos* 99, 457–468 (2002).
[83] P. J. G. Malé, C. Leroy, A. Dejean, A. Quilichini, and J. Orivel, *Evol. Ecol.* 26, 55–63 (2012).
[84] K. E. LeVan, K. L. J. Hung, K. R. McCann, J. T. Ludka, and D. A. Holway, *Oecologia* 174, 163–171 (2014).
[85] T. E. Miller, *J. Ecol.* 102, 496–507 (2014).
[86] S. Koptur, *Ecology* 66, 1639–1650 (1985).
[87] M. Turner, *Funct. Ecol.* 9, 279–284 (1995).
[88] M. Heil, B. Fiala, T. Boller, and K. E. Linsenmair, *Naturwissenschaften* 86, 146–149 (1999).

[89] M. Heil, B. Fiala, B. Baumann, and K. E. *Funct. Ecol.* 14, 749–757 (2000).
[90] L. A. Dyer, C. D. Dodson, J. Beithoffer, and D. K. Letourneau, *J. Chem. Ecol.* 27, 581–592 (2001).
[91] G. Eck, B. Fiala, K. E. Linsenmair, R. B. Hashim, and P. Proksch, *J. Chem. Ecol.* 27, 1979–1996 (2001).
[92] D. K. Letourneau and P. Barbosa, *Biotropica* 31, 295–302 (1999).
[93] D. McKey, *Ame. Nat.* 108, 305–320 (1974).
[94] D. McKey, *The distribution of secondary metabolites within plants. Herbivores: Their Interactions with Secondary Plant Metabolites.* Academic Press, New York pp. 55–133 (1979).
[95] A. Meyer and M. E. Montgomery, *Oecologia* 72, 527–532 (1987).
[96] M. Kielkiewicz and M. van de Vrie, *Exp. Appl. Acarol.* 10, 33–43 (1990).
[97] P. D. Coley and J. A. Barone. *Ann. Rev. Ecol. Syst.* 27, 305–335 (1996).
[98] N. M. van Dam, T. J. de Jong, Y. Iwasa, and T. Kubo, *Funct. Ecol.* 10, 128–136 (1996).
[99] Y. Iwasa, T. Kubo, N. van Dam, and T. J. de Jong, *Theor. Popul. Biol.* 50, 124–148 (1996).
[100] P. W. Lambdon and M. Hassall, *Funct. Ecol.* 19, 299–305 (2005).
[101] V. Radhika, C. Kost, S. Bartram, M. Heil, and W. Boland, *Planta* 228, 449–457 (2008).
[102] N. M. van Dam, R. Verpoorte and E. van der Meijden, *Phytochemistry* 37, 1013–1016 (1994).
[103] L. Z. Meng, K. Martin, J. X. Liu, and J. Chen, *Arth. Plant Interact.* 6, 59–65 (2012).
[104] O. Guhling, C. Kinzler, M. Dreyer, G. Bringmann, and R. Jetter, *J. Chem. Ecol.* 31, 2323–2341 (2005).
[105] S. Mole, Trade-offs and constraints in plant-herbivore defense theory: a life-history perspective. *Oikos* 71, 3–12 (1994).
[106] A. Agrawal, *Trends Ecol. Evol.* 22, 103–109 (2007).
[107] D. McC and H. De Foresta, *Biotropica* 17, 238–244 (1985).
[108] R. A. Wesselingh, *Oecologia* 20, 197–201 (1995).
[109] A. Yamawo, J. Tagawa and N. Suzuki, *Plant Species Biol.* 29, 152–158 (2014).
[110] C. M. de Queiroz, F. V. da Costa, F. de Siqueira Neves, and M. Fagundes, *Arthropod-Plant Interact.* 7, 99–107 (2013).
[111] J. P. Bryant, T. P. Clausen, P. B. Reichardt, M. C. McCarthy, and R. A. *Oecologia* 73, 513–517 (1987).

[112] J. D. C. Hemming and R. L. Lindroth, *J. Chem. Ecol.* 25, 1687–1714 (1999).
[113] J. P. Bryant, F. S. III. Chapin and D. R. Klein, *Oikos* 40, 357–368 (1983).
[114] N. Stamp, *Quarterly Rev. Biol.* 78, 23–55 (2003).
[115] P. J. Folgarait and D. W. Davidson, *Oecologia* 104, 189–206 (1995).
[116] M. Heil, A. Hilpert, B. Fiala, R. Bin Hashim, E. Strohm, G. Zotz, and K. E. Linsenmair, *Functl. Ecol.* 16, 475–483 (2002).
[117] D. Wagner and E. F. Nicklen, *J. Ecol.* 98, 614– 624 (2010).
[118] P. J. Folgarait and D. W. Davidson, *Oikos* 71, 305–320 (1994).
[119] A. Yamawo and Y. Hada, *Ann. Bot.* 106, 143–148 (2010).
[120] R. J. Bixenmann, P. D. Coley and T. A. Kursar, *Oecologia* 165, 417–425 (2011).
[121] M. F. Kersch and C. R. Fonseca, *Ecology* 86, 2117–2126 (2005).
[122] V. Radhika, C. Kost, A. Mithöfer, and W. Boland, *Proc. Natl. Acad. Sci.* 107, 17228–17233 (2010).
[123] A. Yamawo, Y. Hada and N. Suzuki, *J. Plant Res.* 125, 71–76 (2012).
[124] K. Kitajima, A. M. Llorens, C. Stefanescu, M. V. Timchenko, P. W. Lucas, and S. J. Wright, *New Phytol.* 195, 640–652 (2012).
[125] J. W. Dalling and S. P. Hubbell, *J. Ecol.* 90, 557–568 (2002).
[126] S. N. Tripathi and A. S. Raghubanshi, *J. Plant Ecol.* 7, 250–263 (2014).

In: Nectar
Editor: Robin L. Peck
ISBN: 978-1-63463-650-6

Chapter 4

NECTAR PLANT SPECIES RICHNESS EXPLAINS BUTTERFLY DIVERSITY IN AND AROUND A PRIMEVAL WOODLAND OF MOUNT FUJI, CENTRAL JAPAN

Masahiko Kitahara*
Department of Animal Ecology,
Yamanashi Institute of Environmental Sciences (YIES),
Kenmarubi, Fujiyoshida, Yamanashi, Japan

ABSTRACT

To search for factors governing butterfly diversity patterns, we studied adult butterfly communities and their diet resources in and around the Aokigahara primeval woodland of Mount Fuji, central Japan. In case study 1, we established two study sites in the woodland interior, one at the edge of the woodland, and two in the adjacent semi-natural grassland. We monitored adult butterflies and their food items from May to October in 2005 using a transect method. The total number of butterfly species recorded was greatest in the woodland edge site (41 spp.), followed by those in semi-natural grassland sites (39 and 38 spp.). Fewer species were recorded in the woodland interior sites (28 and 20 spp.). This finding supports the commonly held view that butterfly diversity is highest in

* Correspondence: E-mail: mkita@yies.pref.yamanashi.jp.

intermediate successional stages. We also recorded the 38 food items used by adult butterflies. Most were Spermatophyte plants with nectar-bearing flowers. Adult butterflies particularly favored the flowers of perennial herbaceous plants. The total number of butterfly species in each site was significantly positively correlated with the number of herbaceous plant species present. Therefore, the high butterfly species richness in the mid-successional woodland edge and semi-natural grassland habitats appears to be due to the variety of favored food plants in these sites. Seven species mentioned in the Ministry of the Environment Red List were observed during this study. Their populations were predominantly concentrated in the semi-natural grassland sites, suggesting that the preservation of these habitats is important for conserving these endangered butterfly species. In case study 2, we analyzed the relationships between the diversities of vegetation, adult nectar plants, and butterflies in sites in and around the woodland different from those of the case study 1. As well as the results of the case study 1, the nectar resource utilization by adult butterflies deviated extremely to herbaceous plants, especially to perennials, compared to woody species, although most of the study area was in and near the woodland. There were more nectar plant species in sites with more plant species richness. Among the butterfly community indices analyzed, the strongest correlation was detected between butterfly species richness and nectar plant species richness in each site. Another close correlation was detected between the species richness of nectar plants and herbaceous plants in each site. In conclusion, nectar plant species richness is a highly important factor supporting adult butterfly species richness in and around woodland areas. Additionally, the maintenance of herbaceous plant species richness in a butterfly habitat, which leads to the maintenance of nectar plant species richness, is very important for conservation of butterfly diversity even in and around woodland landscape of temperate region.

Keywords: Adult nectar plants, butterfly diversity, conservation value, herbaceous plants, plant-butterfly relations, semi-natural grassland, species richness, vegetation, woodland habitats

INTRODUCTION

To search for the factors governing the diversity patterns of biological communities is one of the central aims of community ecology (MacArthur 1972; Pianka 1988; Begon et al. 1996). This kind of study and information is also vital for the conservation of biodiversity (Primack 1993, 1995). Up to now, butterflies have been well examined as to variation in diversity patterns along a variety of environmental gradients (e.g., Ishii et al. 1991; Spitzer et al. 1993, 1997; Kitahara and Fujii 1994, Blair and Launer 1997; Natuhara et al. 1999; Kitahara et al. 2000; Kocher and Williams 2000; Natuhara 2000; Inoue 2003; Hogsden and Hutchinson 2004). In addition, they have been extensively examined as to their ecological attributes and biotope occupancy (e.g., Dennis et al. 2000; Shreeve et al. 2001; Dennis et al. 2004, 2005), and therefore, repeatedly used in studies of conservation biology as an important bioindicator for environmental assessment (e.g., Sakuratani 1991; Kremen 1992; Schmitt 2003). Thus, butterflies are one of the most suitable organisms for the studies of biological diversity and conservation biology.

In general, as almost all butterflies utilize species-specific plant resources in both larval and adult stages, it is believed that the diversity of plants influences the diversity of butterflies. In fact, the positive correlation between plant and butterfly diversities has been reported or pointed out in many previous studies (e.g., Erhardt 1985; Sparks and Parish 1995; Ishii 1996; Kitahara and Watanabe 2001, 2003; Simonson et al. 2001; Croxton et al. 2005). However, there have been a few studies (Väisänen 1992; Holl 1996; Kitahara 2004) in which the correlation is weak between butterfly diversity and vegetational community composition or species richness. In another study (Hawkins and Porter 2003), it was pointed out that, although plant and butterfly diversities are positively correlated, plant diversity does not directly influence butterfly diversity but that both are probably responding to similar environmental factors. Thus, the actual relationship between butterfly and plant diversities and their causal mechanisms are not yet clear.

On the other hand, it is generally thought that a greater diversity of resources should support a greater diversity of consumers. Indeed, it has been known that a greater abundance and/or diversity of nectar resources are associated with a greater abundance and/or diversity of butterflies (Murphy and Wilcox 1986; Kremen 1992; Munguira and Thomas 1992; Holl 1995; Ishii et al. 1995; Loertscher et al. 1995; Steffan-Dewenter and Tscharntke 1997; Hardy and Dennis 1999; Clausen et al. 2001; Ries et al. 2001; Schneider and Fry 2001; Simonson et al. 2001; Pryke and Samways 2003; Pywell et al. 2004), although an association is absent in a few studies (e.g., Sharp et al. 1974). Schneider and Fry (2001) advocated that the availability of both nectar sources and larval food plants are important in determining butterfly diversity. However, at least in temperate Japan, few studies have been conducted on butterfly species and their nectar plant relationships, and almost no information is available on the conservation value of flowering nectar plants for butterfly abundance and diversity.

Recently, the importance of a resource-founded definition of habitat and approach based on the accurate identification of the spatial and temporal existence of resources in a landscape has been emphasized for butterfly conservation (Dennis et al. 2003, 2006). In addition to this, the need for the development of a resource database on butterfly biology necessary to adopt the resource-based approach has been proposed (Dennis et al. 2003). Thus, in the present study, considering the accumulation of new data on adult nectar resources, we examined the relationships between butterfly species and their nectar plant species in and around Aokigahara primeval woodland and the adjacent semi-natural grassland at the northwestern foot of Mount Fuji, central Japan. Our objectives are (1) to clarify the relationships between the diversities of vegetation, adult nectar plants, and butterfly communities, (2) to examine the role nectar resources play in determining butterfly diversity, and (3) to evaluate the conservation value of flowering adult nectar plants for butterfly diversity.

Case Study 1: Relationship of Butterfly Diversity with Plant Species Richness in and Around A Semi-Natural Grassland of Mount Fuji

In the case study 1, we monitored adult butterfly communities and examined their diet resource utilization in a semi-natural grassland that had been used as meadow in the past, and the adjacent Aokigahara woodland where primeval state is thought to remain at the northwestern foot of Mt. Fuji, central Japan. We established two study sites in the semi-natural grassland (GL-1 and GL-2), two sites in the woodland (FI-1 and FI-2), and one site at the edge of the woodland (FE). In each site, a fixed 300 m-long census route was established. We investigated the communities during the period from May to October in 2005 using the transect method. Details of the study area and methods are described in our previous paper (Kitahara et al., 2010).

1. Total Species Richness and Total Population Density

Figure 1a shows total species richness of butterfly communities during the study period at each study site. The highest species richness (41 spp.) was recognized in the woodland edge site (FE), and followed by in two grassland sites (GL-2 and GL-1). In contrast, the woodland interior sites (FI-1 and FI-2) had few numbers of species, and in particular, the lowest species richness (20 spp.) was recorded in FI-1 site where predominated by conifer trees. These results indicate that the number of adult butterfly species was more in open grassland habitats, and less in closed woodland interior habitats.

Figure 1b shows total population density of butterfly communities during the study period at each study site. The highest total population density (681.5 individuals/ km) was recognized in the semi-natural grassland GL-1 site, and followed by in semi-natural grassland GL-2 site and woodland edge FE site. As well in the trend of total species richness, the woodland interior sites (FI-1 and FI-2) had few numbers of butterfly individuals, and in particular, the lowest density (53.9 individuals/ km) was recorded in FI-1 site where predominated by conifer trees. These results indicate that the number of adult butterfly individuals was more in open grassland habitats, and less in closed woodland interior habitats. This tendency was very clear in density (abundance) more than in species richness.

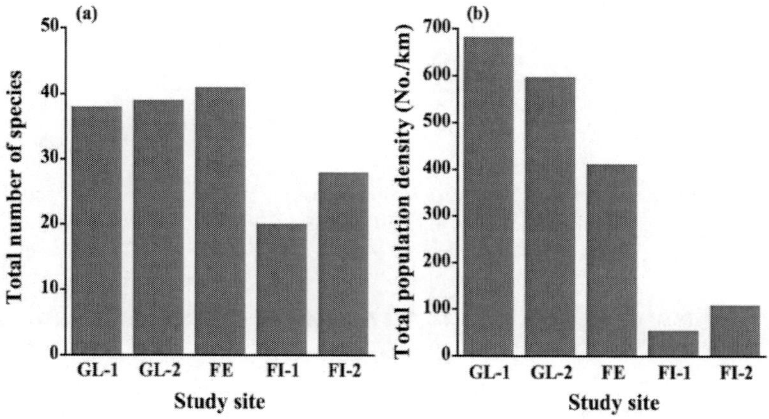

Figure 1. Total number of butterfly species (a) and total butterfly population density (No. individuals /km of transect) (b) in each study site (modified from Kitahara et al. 2010).

2. Adult Diet Resources of Butterflies and Their Utilization Patterns

In this study, we recorded 38 food items (species) used by adult butterflies in all during the study period. Among them, 31 species (81.6%) were Spermatophyte plants with nectar-bearing flowers. The food resources which many species of adult butterflies used were water (sipping mud), *Rubus parvifolius, Cirsium nipponicum, Cirsium japonicum, Ligustrum tschonoskii, Erigeron annuus, Eupatorium lindleyanum, Astilbe microphylla, and Senecio flammeus*, etc. These were almost Spermatophyte plants and butterflies' adults used their nectar. Among the food resources which adult butterflies used, about half of them (20 spp.) was the nectar of perennial herbaceous plants (Figure 2). The next was that of annual herbaceous plants. Contrary to these, the nectar of flowering trees was hardly used for butterflies' adults (ca. 10 %).

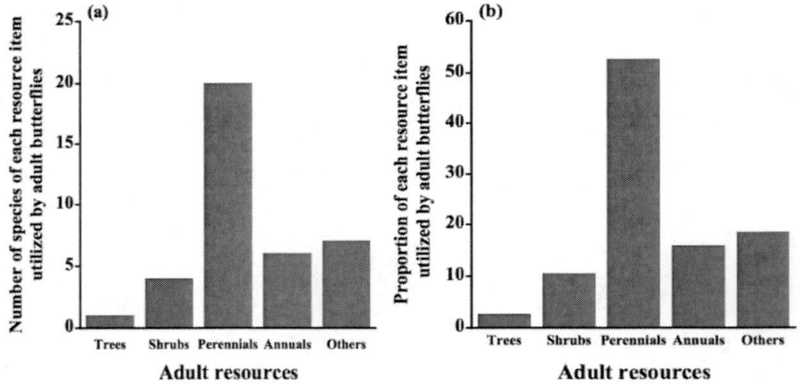

Figure 2. Number of species (a) and proportion (b) of each food plant used by adult butterflies (modified from Kitahara et al. 2010).

3. Relationships between Butterfly and Plant Species Richness

Figure 3 shows the relationships between total number of butterfly species and total number of plant species (a), number of herbaceous plant species (b), and number of woody plant species (c). Among the relationships, the correlation of (b) was only significant. This indicates that butterfly species richness was more influenced by herbaceous plant species richness in the study area.

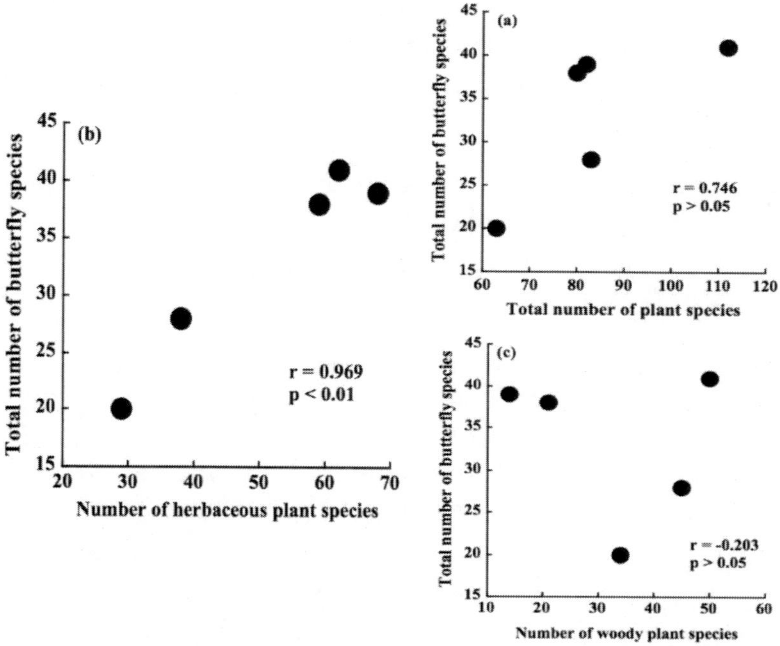

Figure 3. Relationships of total number of butterfly species with total number of plant species (a), number of herbaceous plant species (b), and number of woody plant species (c) (modified from Kitahara et al. 2010).

Case Study 2: Relationship of Butterfly Diversity with Nectar Plant Species Richness in and Around the Aokigahara Primeval Woodland of Mount Fuji

In the case study 2, we examined the relationships between the diversities of vegetation, adult nectar plants, and butterflies in and around the Aokigahara primeval woodland (forest) on the northwestern footslopes of Mount Fuji, central Japan. We selected six study sites comprising three habitat types: forest interior (Sites FI-1 and FI-2), forest edge (Sites FE-1 and FE-2), and open land (Sites OL-1 and OL-2) adjacent to the forest. All study sites were similar in terms of altitude (ca. 1000 m a.s.l.) and topography (almost flat or gently sloping land), and were located inside an area measuring 2.4 km east to west and 0.63 km north to south. In each site, a fixed 300 m-long census route was established. We investigated the communities and the vegetation during the period from April to November in 1999 using the transect method. Details of the study area and methods are described in our previous paper (Kitahara et al., 2008).

1. Vegetation and Adult Nectar Plants

The values of various indices of butterfly communities, vegetation represented by plants belonging to Spermatophyta, and nectar plants utilized by adult butterflies in each of all the six study sites are shown in Table 1.

Table 1. The values of various indices of butterfly communities, vegetation, and nectar plants recorded in the six study sites. (modified from Kitahara et al. 2008)

Study site	OL-1	OL-2	FE-1	FE-2	FI-1	FI-2
Butterfly community						
Total population density	74.76	121.87	269.53	162.92	10.83	9.44
Total number of species	18	23	43	39	3	3
Species diversity (*H'*)	2.633	2.659	3.240	3.364	1.058	1.028
Species diversity (1- λ)	0.920	0.900	0.930	0.960	0.700	0.700
Vegetation (Spermatophyta)						
No. all plant species	52	60	106	136	37	64
No. woody plant species	1	16	35	55	34	42
No. tree species	0	6	16	31	20	22
No. shrub species	1	9	15	20	13	17
No. herbaceous plant species	50	42	69	77	2	21
No. perennial species	23	24	48	55	2	17
No. annual species	27	18	18	20	0	3
Adult nectar plants						
No. all nectar plant species	17	19	28	28	3	8
No. woody nectar plant species	1	2	3	4	3	4
No. tree nectar species	0	1	1	1	2	2
No. shrub nectar species	1	1	2	3	1	2
No. herbaceous nectar plant species	16	17	25	24	0	4
No. perennial nectar species	11	12	17	16	0	3
No. annual nectar species	5	5	8	8	0	1

Table 2. Number of species of plants belonging to Spermatophyta, and the number of nectar plant species among them recorded across all study sites. (modified from Kitahara et al. 2008)

	Herbaceous plants			Woody plants			Other plants	Total
	Annuals	Perennials	Total	Shrubs	Trees	Total		
Plants belonging to Spermatophyta	43	84	127	40	44	84	10	221
Adult nectar plants	11	22	33	3	2	5	0	38

Table 2 shows the species composition of all plants (belonging to Spermatophyta) and adult nectar plants recorded all over the study sites. Out of all 221 plant species (Spermatophyta) recorded in this study, their utilization as a nectar resource by adult butterflies was observed in 38 plant species (17.2 %). Of all 127 herbaceous and 84 woody species, the proportion used as a nectar resource by adult butterflies was substantially higher in herbaceous (33 spp., 26.0 %) than woody (6 spp., 5.9 %) plants. This trend is shown in Figure 4.

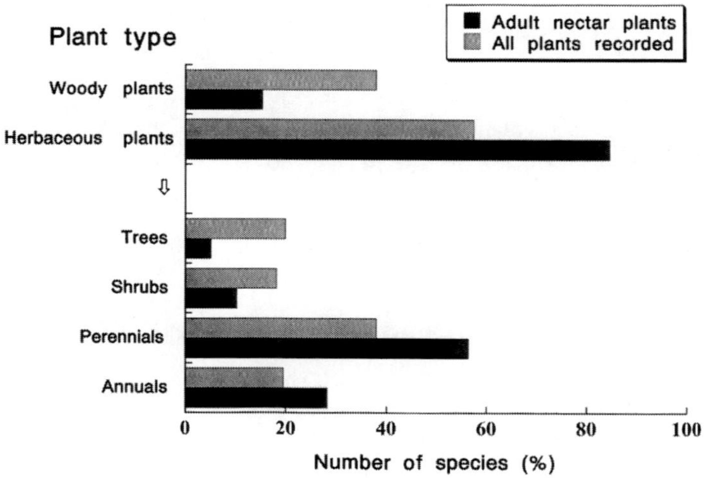

Figure 4. The number of species (%) of the respective plant types in all plants belonging to Spermatophyta, and in adult nectar plants among them recorded in all the study area (modified from Kitahara et al. 2008).

The chi-square test for goodness of fit showed that the proportions of number in the woody and herbaceous plant species recorded in the study area (i.e., expected proportions) differed significantly from those of number of woody and herbaceous adult nectar plant species used (i.e., observed proportions) ($\chi^2 = 26.333$, df = 1, $P < 0.0001$). In more detailed analysis, the chi-square test for goodness of fit showed that the proportions of number in the species of trees, shrubs, perennials, and annuals recorded in the study area (i.e., expected proportions) differed significantly from those of number of adult nectar species of trees, shrubs, perennials, and annuals used (i.e., observed proportions) ($\chi^2 = 27.246$, df = 3, $P < 0.0001$). These results indicate that, in the study area, the nectar resource utilization by adult butterflies is biased heavily to herbaceous plants especially perennials, compared to woody plants.

A significant positive correlation in the number of species was detected between adult nectar plants and all, herbaceous, and perennial plants belonging to Spermatophyta recorded at each study site (Table 3). In particular, the correlation was strong in both herbaceous and perennial plants.

Table 3. Correlation coefficients of the number of species between various plant types and adult nectar plants recorded in each study site. (modified from Kitahara et al. 2008)

	All plants	Herbaceous plants	Annual plants	Perennial plants	Woody plants	Tree plants	Shrub plants
Nectar plants	0.843*	0.983***	0.767	0.958***	0.140	0.061	0.161

*** $P < 0.001$

* $P < 0.05$

2. Butterfly Communities and Adult Nectar Plants

All butterfly community indices (total population density, total number of species, species diversities *H'* and 1-λ) were positively and highly significantly correlated with the number of species of all adult nectar plants recorded at each study site. In particular, the strongest correlation was detected with butterfly species richness (total no. of species) (*Table 4*). On the other hand, similar to the above results, all butterfly community indices were positively and highly significantly correlated with the number of species of herbaceous adult nectar plants recorded in each study site.

Table 4. Correlation coefficients between various butterfly community indices and the numbers of nectar plant species in the six study sites (modified from Kitahara et al. 2008)

	Total population density	Butterfly community index Total number of species	Species diversity (H')	Species diversity (1-λ)
All nectar plants	0.915**	0.979***	0.970***	0.926***
Herbaceous nectar plants	0.907**	0.968***	0.986***	0.958***
Woody nectar plants	0.016	0.045	-0.198	-0.328

*** $P < 0.001$.
** $P < 0.01$.
* $P < 0.05$.

However, no significant correlations were detected between all butterfly community indices and the number of species of woody adult nectar plants recorded at each study site.

DISCUSSION

1. Resource Utilization Patterns of Adult Butterflies

In two case studies, we have shown that the nectar resource utilization by adult butterflies is heavily biased to herbaceous plants, especially to perennials, compared to woody species, although most of the study areas were in and near a primeval woodland. Such biased herb resource utilization by adult butterflies has been observed by several previous authors (Kitahara 2000; Kamimura 2004; Mano 2004; Tiple et al. 2006; Nagata et al. 2007). In contrast, biased nectar use of woody plant species by adult butterflies has hardly been reported. Also, in arable field margins of Britain, the importance of perennial nectar sources rather than annual ones was pointed out for butterfly conservation (Dover 1996). In these studies, the highest number of adult butterfly species was recorded in woodland edge study sites, intermediate in open land sites, and lowest in woodland interior sites (cf., Kitahara and Watanabe 2003). It is also a general observation that most adult butterflies avoid shade and are often encountered in open sunny places (Douwes 1975, Dennis and Bramley 1985; Warren 1985; Pivnick and McNeil

1987). Thus, one possible reason for biased herb resource use by adults is that most adult butterflies indeed prefer flowers of herbaceous plants to those of woody ones, or that herb abundance and density are simply much higher in open sunny spaces such as woodland edges and open land sites with abundant adult butterflies. Concerning butterfly conservation, the biased herb resource use by adults suggests that the maintenance of herbaceous plant species richness and diversity in their habitats is important for ensuring the nectar resources of adult butterflies.

2. Relationship between Vegetation and Adult Nectar Plants

In case study 2, the number of species of adult nectar plants in each site was significantly and positively correlated with that of all plants belonging to Spermatophyta. Thus, there were greater nectar plant species in sites with greater plant species richness. In more detail, the number of species of adult nectar plants at each site was more strongly correlated with that of herbaceous or perennial plants than with that of all plants belonging to Spermatophyta, suggesting that the numbers of species of herbaceous and perennial plants are both good predictors for adult nectar plants at the respective study sites. Probably, the strong correlations in numbers of species between adult nectar plants and both herbaceous and perennial plants at each site are caused by bias in herb resource utilization of adult butterflies. The relationships between vegetation and adult nectar plants suggest that the maintenance and management of habitats for species richness and diversity of herbaceous plants is important for the supply and availability of adult nectar resources.

3. Relationship between Butterfly Diversity and Adult Nectar Plants, and Conservation Implications

In case study 1, butterfly species richness was positively and significantly correlated with number of herbaceous plant species in each site, but not with total number of plant species (Figure 3). Whilst, the nectar resource use by adult butterflies was heavily biased to herbaceous plant species. This suggests that butterfly species richness was strongly governed by the abundance of adult food resources in each site. That butterfly diversity patterns can be explained well in terms of food resource abundance has been reported in other previous studies (Kitahara and Watanabe 2003; Yamamoto et al. 2007). Thus, it is concluded that food resource abundance is one of important factors affecting butterfly diversity patterns.

In case study 2, we showed that butterfly community indices were all positively correlated with the number of nectar plant species at each site. Among them, the strongest correlation was detected between butterfly species richness (total no. of species) and nectar plant species richness ($r = 0.979$, $p < 0.001$). This correlation with the number of species of nectar plants is much stronger than that with the number of species of all plants (belonging to Spermatophyta) recorded at each site ($r = 0.842$, $p < 0.05$) shown in a previous study (Kitahara and Watanabe 2001). These results suggest that nectar plant species richness is an important factor governing adult butterfly species richness at each site rather than total species richness of plants (Figure 5). The importance of nectar plant abundance for butterfly abundance and diversity has been pointed out in many previous studies (Murphy and Wilcox 1986; Kremen 1992; Munguira and Thomas 1992; Holl 1995; Ishii et al. 1995; Loertscher et al. 1995; Steffan-Dewenter and Tscharntke 1997; Hardy and Dennis 1999; Clausen et al. 2001; Ries et al. 2001; Schneider and Fry 2001; Simonson et al. 2001; Pryke et al. 2003; Pywell et al. 2004), although the reverse has emerged in a few studies (Sharp et al. 1974). A number of population and community studies of butterflies suggest that adult distribution patterns are more affected by the availability of nectar resources than the presence of larval host plants (Ehrlich and Gilbert 1973; Gilbert and Singer 1973; Murphy 1983; Grossmueller 1987; Feber et al. 1996; Hardy and Dennis 1999).

In addition, it has been known in some studies (Thomas and Mallorie 1985; Holl 1996) that butterfly species richness is positively correlated with herbaceous cover. Our studies also showed that butterfly community indices were all strongly correlated with the species richness of herbaceous nectar plants at each site, but not with that of woody nectar plants. This indicates that the richness of herbaceous nectar plant species is one of the most important factors determining the community structure and attributes of adult butterflies. Yet, most of the present study area was in and near a primeval woodland; nevertheless, the richness of woody nectar plant species has almost no effects on the determination of adult butterfly community structure.

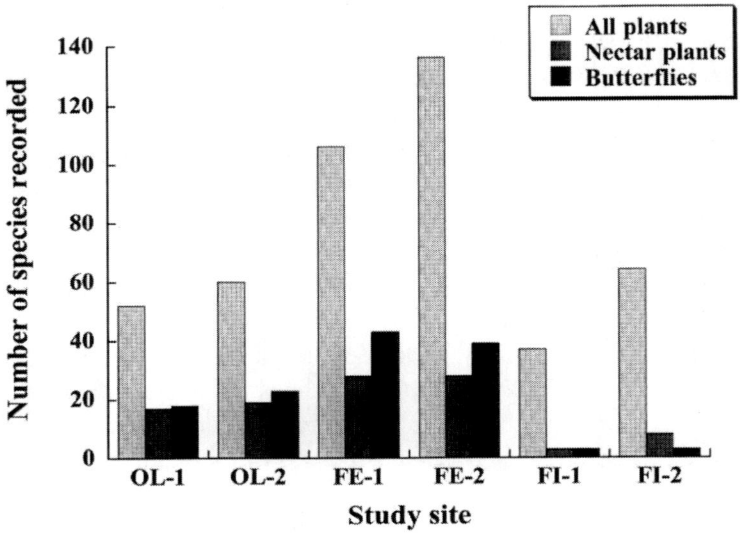

Figure 5. Number of species of all plants belonging to Spermatophyta, adult nectar plants, and butterflies recorded in each study site (modified from Kitahara et al. 2008).

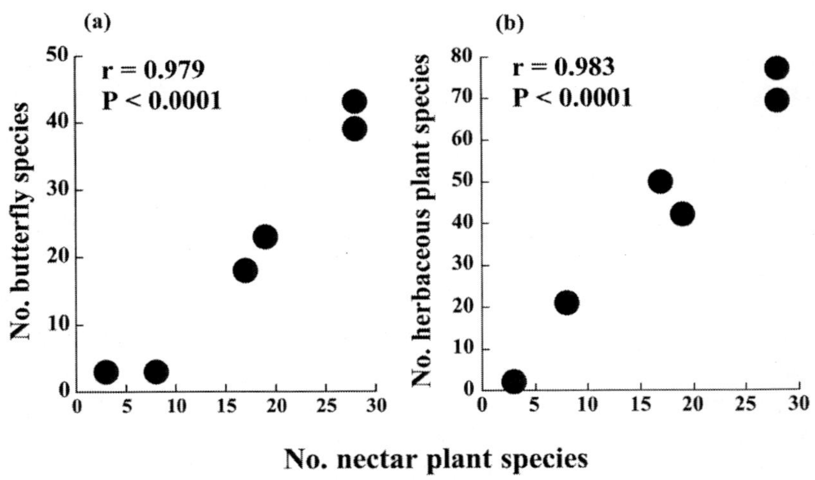

Figure 6. Relationships between the number of nectar plant species and the number of butterfly species (a), and the number of herbaceous plant species (b) recorded in each study site (modified from Kitahara et al. 2008).

The importance of herbaceous plant species richness even within woodlands for herbivore diversity is also known for moths (Usher and Keiller

1998). In case study 2, two key correlations are found, one between the species richness of butterflies and nectar plants, and another between the species richness of nectar plants and herbaceous plants at each site (Figure 6).

In conclusion, herbaceous plant species richness in a habitat has a central role in its nectar plant species richness, and that nectar plant richness is a highly important factor governing and supporting its adult butterfly species richness. The maintenance and management of herbaceous plant species richness in a butterfly habitat, which underlies nectar plant species richness, is very important for the maintenance and conservation of butterfly species richness and diversity even in and around woodland landscapes of temperate region, as Tudor et al. (2004) claim that management of woodland sites for butterfly conservation should give as much consideration to nectar sources as to host plant sources.

ACKNOWLEDGMENTS

I am grateful to Dr. T. Kobayashi, Ms. M. Yumoto, Ms. A. Kakizaki, Dr. T. Nakano, Dr. E. Maruta and Dr. T. Yasuda who are co-investigators of the present study. My thanks are also due to Drs. H. Imaki, Z. Jiang, H. Ueda, Y. Yoshida and Ms. A. Kobayashi of the Lab. of Animal Ecology, and other researchers of the Yamanashi Institute of Environmental Sciences for their discussion, help and understanding of this study. I am deeply indebted to Prof. Emeritus K. Fujii of the Univ. of Tsukuba, and Prof. Emeritus T. Nakamura of the Univ. of Yamanashi for their continuous advice and encouragement throughout the study. This work was supported in part by both Grants-in-Aid for Scientific Research (B) (no. 17310138) and for Scientific Research (C) (no. 20510221) from the Japan Society for the Promotion of Science (JSPS) to M. Kitahara.

REFERENCES

Ae, SA, Hirowatari T, Ishii, M, Brower, LP (eds.) (1996) Decline and Conservation of Butterflies in Japan Ⅲ. The Lepidopterological Society of Japan, Osaka, Japan.

Begon M, Harper JL, Townsend CR (1996) Ecology: individuals, populations, and communities, 3rd edn. Blackwell Science, Oxford.

Blair RB, Launer AE (1997) Butterfly diversity and human land use: species assemblages along an urban gradient. *Biol. Conserv.* 80:113-125.

Clausen HD, Holbeck HB, Reddersen J (2001) Factors influencing abundance of butterflies and burnet moths in the uncultivated habitats of an organic farm in Denmark. *Biol. Conserv.* 98:167-178.

Cook LM, Dennis RLH, Hardy PB (2001) Butterfly-hostplant fidelity, vagrancy and measuring mobility from distribution maps. *Ecography* 24:497-504.

Croxton PJ, Hann JP, Greatorex-Davies JN, Sparks TH (2005) Linear hotspots? The floral and butterfly diversity of green lanes. *Biol. Conserv.* 121:579-584.

Dennis RLH (2001) Progressive bias in species status in symptomatic of fine-grained mapping units subject to repeated sampling. *Biodivers. Conserv.* 10:483-494.

Dennis RLH (2004) Butterfly habitats, broad-scale biotope affiliations, and structural exploitation of vegetation at finer scales: the matrix revisited. *Ecol. Entomol.* 29:744-752.

Dennis RLH, Bramley MJ (1985) The influence of man and climate on dispersion patterns within a population of adult *Lasiommata megera* (L.) (Satyridae) at Brereton Hearth Cheshire (U.K.). *Nota Lepid* 8:309-324.

Dennis RLH, Donato B, Sparks TH, Pollard E (2000) Ecological correlates of island incidence and geographical range among British butterflies. *Biodivers. Conserv.* 9:343-359.

Dennis RLH, Hodgson JG, Grenyer R, Shreeve TG, Roy DB (2004) Host plants and butterfly biology: Do host-plant strategies drive butterfly status? *Ecol. Entomol.* 29:12-26.

Dennis RLH, Shreeve TG, Arnold HR, Roy DB (2005) Does diet breadth control herbivorous insect distribution size? Life history and resource outlets for specialist butterflies. *J. Insect. Conserv.* 9:187-200.

Dennis RLH, Shreeve TG, Van Dyck H (2006) Habitats and resources: the need for a resource-based definition to conserve butterflies. *Biodivers. Conserv.* 15:1943-1966.

Dennis RLH, Shreeve TG, Van Dyck H (2003) Towards a functional resource-based concept for habitat: a butterfly biology viewpoint. Oikos 102:417-426.

Dennis RLH, Sparks TH, Hardy PB (1999) Bias in butterfly distribution maps: the effects of sampling effort. *J. Insect. Conserv.* 3:33-42.

Douwes P (1975) Distribution of a population of the butterfly *Heodes virgaureae*. Oikos 26:332-340.

Dover JW (1996) Factors affecting the distribution of satyrid butterflies on arable farmland. *J. Appl. Ecol.* 33:723-734.

Ehrlich PR, Gilbert LE (1973) Population structure and dynamics of the tropical butterfly *Heliconius ethilla*. *Biotropica* 5:69-82.

Erhardt A (1985) Diurnal Lepidoptera: sensitive indicators of cultivated and abandoned grassland. *J. Appl. Ecol.* 22:849-861.

Feber RE, Smith H, Macdonald DW (1996) The effects on butterfly abundance of the management of uncropped edges of arable fields. *J. Appl. Ecol.* 33:1191-1205.

Gall LF (1985) Measuring the size of Lepidopteran populations. J Res Lepid 24:97-116

Gilbert LE, Singer MC (1973) Dispersal and gene flow in a butterfly species. *Amer. Natur.* 107:58-72.

Grossmueller DW (1987) The role of nectar source distribution in habitat use and oviposition by the tiger swallowtail butterfly. *J. Lepid. Soc.* 41:159-165.

Hardy PB, Dennis RLH (1999) The impact of urban development on butterflies within a city region. *Biodivers. Conserv.* 8:1261-1279.

Hawkins BA, Porter EE (2003) Does herbivore diversity depend on plant diversity? The case of California butterflies. *Amer. Natur.* 161:40-49.

Hogsden KL, Hutchinson TC (2004) Butterfly assemblages along a human disturbance gradient in Ontario, Canada. *Can. J. Zool.* 82:739-748.

Holl KD (1995) Nectar resources and their influence on butterfly communities on reclaimed coal surface mines. *Restoration Ecol.* 3:76-85.

Holl KD (1996) The effect of coal surface mine reclamation on diurnal lepidopteran conservation. *J. Appl. Ecol.* 33:225-236.

Inoue T (2003) Chronosequential change in a butterfly community after clear-cutting of deciduous forests in a cool temperate region of central Japan. *Entomol. Sci.* 6:151-163.

Ishii M (1996) Species diversity of butterfly communities in different environment of forests in southern Osaka. In: Tanaka B, Arita Y (eds) Decline and conservation of butterflies in Japan, vol IV. The Lepidopterological Society of Japan, Osaka, pp 63-75 (in Japanese with English summary).

Ishii M, Hirowatari T, Fujiwara S (1995) Species diversity of butterfly communities in 'Mt, Mikusa Coppice' for Zephyrus. *Jpn. J. Environ. Entomol. Zool.* 7:134-146 (in Japanese with English summary).

Ishii M, Yamada M, Hirowatari T, Yasuda T (1991) Diversity of butterfly communities in urban parks in Osaka prefecture. *Jpn. J. Environ. Entomol.* Zool 3:183-195 (in Japanese with English summary).

Kamimura Y (2004) Nectar-source plants and their usage patterns by an adult butterfly community in a riparian biotope. *Natur. Environ. Sci. Res.* 17:107-115 (in Japanese with English summary).

Kitahara M (2000) Food resource usage patterns of adult butterfly communities in woodland habitats at the northern foot of Mt. Fuji, central Japan. *Jpn. J. Environ. Entmol. Zool.* 11:61-81 (in Japanese with English summary).

Kitahara M (2004) Butterfly community composition and conservation in and around a primary woodland of Mount Fuji, central Japan. *Biodivers. Conserv.* 13:917-942.

Kitahara M, Fujii K (1994) Biodiversity and community structure of temperate butterfly species within a gradient of human disturbance: an analysis based on the concept of generalist vs. specialist strategies. *Res. Popul. Ecol. 36*:187-199.

Kitahara M, Kakizaki A, Nakano T, Maruta E, Yasuda T (2010) Conservation value of semi-natural grassland habitats for butterfly diversity and their endangered species in the northwestern area of Mount Fuji, central Japan. *Jpn. J. Environ. Entomol. Zool.* 21:115-125 (in Japanese with English abstract).

Kitahara M, Sei K, Fujii K (2000) Patterns in the structure of grassland butterfly communities along a gradient of human disturbance: further analysis based on the generalist/specialist concept. *Popul. Ecol.* 42:135-144.

Kitahara M, Watanabe M (2001) Relationships of butterfly community diversity to vegetational species richness in and around the Aokigahara woodland at the northern foot of Mt. Fuji, central Japan. *Jpn. J. Environ. Entomol. Zool.* 12:131-145 (in Japanese with English abstract).

Kitahara M, Watanabe M (2003) Diversity and rarity hotspots and conservation of butterfly communities in and around the Aokigahara woodland of Mount Fuji, central Japan. *Ecol. Res.* 18:503-522.

Kitahara M, Yumoto M, Kobayashi T (2008) Relationship of butterfly diversity with nectar plant species richness in and around the Aokigahara primary woodland of Mount Fuji, central Japan. *Biodiv. Conserv.* 17: 2713-2734.

Kocher SD, Williams EH (2000) The diversity and abundance of North American butterflies vary with habitat disturbance and geography. *J. Biogeogr.* 27:785-794.

Kremen C (1992) Assessing the indicator properties of species assemblages for natural areas monitoring. *Ecol. Appl.* 2: 203-217.

Lincoln R, Boxshall G, Clark P (1998) A dictionary of ecology, evolution and systematics, 2nd edn. Cambridge University Press, Cambridge.

Loertscher M, Erhardt A, Zettel J (1995) Microdistribution of butterflies in a mosaic-like habitat: the role of nectar sources. *Ecography* 18:15-26.

MacArthur RH (1972) Geographical ecology: patterns in the distribution of species. Harper & Row, New York.

Mano T (2004) Butterfly communities in the center of Toyota City. Yahagi-River Research 8:115-121 (in Japanese).

Munguira ML, Thomas JA (1992) Use of road verges by butterfly and burnet populations, and the effect of roads on adult dispersal and mortality. *J. Appl. Ecol.* 29:316-329.

Murphy DD (1983) Nectar sources as constraints on the distribution of egg masses by the checkerspot butterfly, *Euphydryas chalcedona* (Lepidoptera: Nympharidae). *Environ. Entomol.* 12:463-466.

Murphy DD, Wilcox BA (1986) Butterfly diversity in natural habitat fragments: a test of the validity of vertebrate-based management. In:Verner J, Morrison M, Ralph CJ (eds) Wildlife 2000, modeling habitat relationships of terrestrial vertebrates. University of Wisconsin Press, Madison, pp 287-292.

Nagata N, Iizuka H, Kitahara M (2007) Food resource usage patterns of adult butterfly communities in the mountainous area of Iwaki City, Fukushima, Japan. *Jpn. J. Environ. Entomol. Zool.* 18:153-160 (in Japanese with English abstract).

Natuhara Y (2000) Changes in butterfly assemblage along the urban-forest gradient. *Jpn. J. Landscape Res.* 63:515-518 (in Japanese with English summary).

Natuhara Y, Imai C, Takahashi M (1999) Pattern of land mosaics affecting butterfly assemblage at Mt. Ikoma, Osaka, Japan. *Ecol. Res.* 14:105-118.

New TR (1991, 1997) Butterfly conservation, 1st edn, 2nd edn. Oxford University Press, Melbourne, Australia.

Pianka ER (1988) Evolutionary ecology, 4th edn. Harper & Row, New York.

Pivnick KA , McNeil JN (1987) Diel patterns of activity of *Thymelicus lineola* adults (Lepidoptera:Hesperiidae) in relation to weather. *Ecol. Entomol.* 12:197-207.

Pollard E (1977) A method for assessing changes in the abundance of butterflies. *Biol. Conserv.* 12:115-134.

Pollard E (1984) Synoptic studies on butterfly abundance. In: Vane-Wright RI, Ackery PK (eds) The Biology of Butterflies. Academic Press, London, pp 59-61.

Pollard E, Yates TJ (1993) Monitoring butterflies for ecology and conservation. Chapman & Hall, London.

Primack RB (1993) Essentials of conservation biology. Sinauer Associates, Inc.,Sunderland, Massachusetts.

Primack RB (1995) A primer of conservation biology. Sinauer Associates, Inc.,Sunderland, Massachusetts.

Pryke SR, Samways MJ (2003) Quality of remnant indigenous grassland linkages for adult butterflies (Lepidoptera) in an afforested African landscape. *Biodivers. Conserv.* 12:1985-2004.

Pywell RF, Warman EA, Sparks TH, Greatorex-Davies JN, Walker KJ, Meek WR, Carvell C, Petit S, Firbank LG (2004) Assessing habitat quality for butterflies on intensively managed arable farmland. *Biol. Conserv.* 118:313-325.

Ries L, Debinski DM, Wieland ML (2001) Conservation value of roadside prairie restoration to butterfly communities. *Conserv. Biol.* 15:401-411.

Sakuratani Y, Fujiyama S (1991) Influence of highway construction on butterfly communities. *Jpn. J. Environ. Entomol. Zool.* 3:15-23 (in Japanese with English summary).

Schmitt T (2003) Influence of forest and grassland management on the diversity and conservation of butterflies and burnet moths (Lepidoptera, Papilionoidea, Hesperiidae, Zygaenidae). *Anim. Biodivers. Conserv.* 26:51-67.

Schneider C, Fry GLA (2001) The influence of landscape grain size on butterfly diversity in grasslands. *J. Insect Conserv.* 5:163-171.

Seido K (1991) Age of the hinoki (*Chamaecyparis obtusa*) natural stands and their distribution on the Aokigahara Woodland, Mt. Fuji. Transactions of the Meeting in Kanto Branch of the Japanese Forestry Society 42:33-36 (in Japanese).

Sharp MA, Parks DR, Ehrlich PR (1974) Plant resources and butterfly habitat selection. *Ecology* 55:870-875.

Shreeve TG, Dennis RLH, Roy DB, Moss D (2001) An ecological classificationof British butterflies: ecological attributes and biotope occupancy. *J. Insect. Conserv.* 5:145-161.

Shreeve TG, Mason CF (1980) The number of butterfly species in woodlands. *Oecologia* 45:414-418.

Simonson SE, Opler PA, Stohlgren TJ, Chong GW (2001) Rapid assessment of butterfly diversity in a montane landscape. *Biodivers. Conserv.* 10:1369-1386.

Simpson EH (1949) Measurement of diversity. *Nature* 163:688.

Sparks TH, Parish T (1995) Factors affecting the abundance of butterflies in field boundaries in Swavesey Fens, Cambridgeshire, UK. *Biol. Conserv.* 73:221-227.

Spitzer K, Jaros J, Havelka J, Leps J (1997) Effect of small-scale disturbance on butterfly communities of an Indochinese montane rainforest. *Biol. Conserv.* 80:9-15

Spitzer K, Novotny V, Tonner M, Leps J (1993) Habitat preferences, distribution and seasonality of the butterflies (Lepidoptera, Papilionoidea) in a montane tropical rain forest, Vietnam. *J. Biogeogr.* 20:109-121.

Steffan-Dewenter I, Tscharntke T (1997) Early succession of butterfly and plant communities on set-aside fields. *Oecologia* 109:294-302.

Thomas CD, Mallorie HC (1985) Rarity, species richness and conservation: butterflies of the Atlas Mountains in Morocco. *Biol. Conserv.* 33:95-117.

Thomas JA (1983) A quick method for estimating butterfly numbers during surveys. *Biol. Conserv.* 27:195-211.

Tiple AD, Deshmukh VP, Dennis RLH (2006) Factors influencing nectar plant resource visits by butterflies on a university campus: implications for conservation. *Nota lepid* 28:213-224.

Tudor O, Dennis RLH, Greatorex-Davies JN, Sparks TH (2004) Flower preferences of woodland butterflies in the UK: nectaring specialists are species of conservation concern. *Biol. Conserv.* 119:397-403.

Usher MB, Keiller SWJ (1998) The macrolepidoptera of farm woodlands: determinants of diversity and community structure. *Biodivers. Conserv.* 7:725-748.

Väisänen R (1992) Distribution and abundance of diurnal Lepidoptera on a raised bog in southern Finland. *Annal. Zool. Fenn.* 29:75-92.

Warren MS (1985) The influence of shade on butterfly numbers in woodland rides, with special reference to the wood white *Leptidea sinapis*. *Biol. Conserv.* 33:147-164.

Warren MS (1985) The influence of shade on butterfly numbers in woodland rides, with special reference to the wood white *Leptidea sinapis*. *Biol. Conserv. 33*:147-164.

Yamamoto N, Yokoyama J, Kawata M (2007) Relative resource abundance explains butterfly biodiversity in island communities. *Proc. Nat. Acad. Sci. USA* 104:10524-10529.

In: Nectar
Editor: Robin L. Peck
ISBN: 978-1-63463-650-6

Chapter 5

SOME MELLIFEROUS ASPECTS OF COHABITING LAMIACEAE SPECIES COMPETING FOR EFFICIENT POLLINATORS

Marina Mačukanović-Jocić[1,*], Gojko Jocić[2], Zora Dajić Stevanović[1] and Mića Mladenović[1]
[1]University of Belgrade, Faculty of Agriculture, Belgrade, Serbia
[2]"Karadjordje" Elementary School, Belgrade, Serbia

ABSTRACT

Some flower morphophysiological and pollinator-luring features as well as co-adaptive traits, including nectar production, nectaries and pollen of five Lamiaceae species, namely *Hyssopus officinalis, Lavandula officinalis, Melissa officinalis, Salvia officinalis* and *Salvia sclarea*, were studied, with reference to the pollinator visits, especially considering honey bee forage preferences. Although these species are cultivated in plantation for medicinal purposes, at the same time they represent a significant source of pollen and nectar for honey production, therefore the research aimed to estimate their melliferous potential, attractiveness and significance for bee pasture.

The pollen grains are isopolar, radially symmetric, predominantly hexacolpate in all species, but often mixed together with tetracolpate and

* Corresponding author: Marina Mačukanović-Jocić. University of Belgrade, Faculty of Agriculture, Belgrade, Serbia. E-mail: marmajo@agrif.bg.ac.rs.

octacolpate in *S. sclarea*. They are mostly medium-sized to large. The smallest grain was found in *L. officinalis* and the largest one in *Salvia sclarea*. The exine sculpturing is bireticulate (*H. officinalis, M. officinalis, S. officinalis*), reticulate (to foveolate) (*S. sclarea*) and foveolate (to perforate) (*L. officinalis*). The shape of pollen grains in equatorial view varies among different species, and within the same species. In the majority of studied species, shape is most frequently prolate (*L. officinalis* - type a, *M. officinalis, S. officinalis* and *S. sclarea* - Class I), except in *H. officinalis* and *S. sclarea* - Class II, in which they are sub-prolate and *L. officinalis* - type b in which they are oblate-spheroidal.

Floral nectaries are located on the receptacle, surrounding the ovary base in the shape of an asymmetrical annular disc with four lobes. Unevenly distributed modified stomata were observed on the upper part or the entire surface of the functional, and generally, the most developed nectary lobe facing the lower corolla lip. The nectariferous tissue is composed of the uniseriate epidermis overlying several layers of polygonal cells forming subepidermal secretory tissue. Epidermal cells are lined with very thin cuticle. Xylem elements alone, separating from the vasculature of the flower receptacle, were noticeable nearby or inside the nectariferous tissue.

The highest total daily and summary diurnal nectar amount per flower, as well as the nectar secretion rate were found in *S. sclarea,* and the lowest in the M. *officinalis* respectively. Pattern of diurnal dynamics of nectar secretion with one secretion peak was observed: in the morning, in both *Salvia* species, *H. officinalis* and *L. officinalis,* and early in the afternoon in *M officinalis.*

Pollinator observation indicated the visitation of seven species of Hymenoptera, three species belonging to Diptera and two species of daily-active Lepidoptera. The honey bee (*Apis mellifera*) was the most abundant pollinator. The major elements of attractiveness were flower scent and colour, followed by the shape and size of the corolla and to a lesser extent the nectar amount.

According to all analyzed flower characteristics and observation of pollinator behaviour and visits, the most attractive plants were *H. officinalis, S. officinalis* and *L. officinalis*, whereas *M. officinalis* was the least attractive. *S. sclarea* was completely unattractive to all recorded pollinators, except *Xylocopa violacea*.

Keywords: Lamiaceae, honey bee, pollinator, nectar production, flower morphology

INTRODUCTION

The appearance of the flowering plants from the evolutionary point of view was responsible for the emergence of certain animals committed to them and specialised for their pollination. Their evolution has ever since been taking its course in parallel, through co-adaptation. By becoming adapted to one another they had entered into a mutually beneficial relationship which implied great evolutionary pressure on the plants, and forced them to constantly search for new options to be as attractive as possible and to succeed in ruthless competition with other plants for the most reliable pollinator. In their quest for a reliable partner, the plants skilfully "chose" their suitors, attracting some, and repelling others. This is how the zoophilous plants over millions of years acquired floral adaptive characteristics that enabled them to be efficiently pollinated by a certain class or species of animal.

On the other hand, pollinators also acquired adaptive characteristics, typical for their type of feeding, which led them to become specialized in the pollination of a particular plant family, genus or even a species. This co-evolutionary race and plant-pollinator circumvention that lasted millions of years, is still ongoing for many competitive couples.

The necessity of mutual assistance arose from the tendency of the flowering plant and the pollinator to perfect their reproductive and survival strategies in an ever competitive environment. In this interdependent relationship, the two partners have different interests but are striving for the same goal. The plant aims to be efficiently pollinated, and the animal seeks to find food in the quickest and easiest way, and both partner's final aim is the survival and the extension of their own species.

Certain groups of pollinators are lured to visit flowers by a wide variety of attractants, especially nectar. The unique combination of flower traits such as odour, colour, shape, size, nectar amount and composition, pollen, honey guides, etc., is responsible for making them attractive to a particular type of pollinator (Faegri and van der Pijl, 1979; Proctor et al., 1996). For being successfully pollinated, flowers recompense visitors for their service by offering them products enriched in sugar, amino acids and other nutrients.

Therefore, this paper shall study the wonderful world of melliferous plants and pollinators. The study of this topic opens many questions, such as: What are these plants we call melliferous? Has their occurrence been justified from the evolutionary point of view? What is their purpose? Why are they significant? Whom do they attract? What do they use for attracting? Is it possible to assess their melliferous potential? etc.

Even though their name seems to indicate so, this group of plants cannot be defined only from the point of view of their importance for the honey bee. Perhaps the most acceptable definition of melliferous plans is that these are flowering plants whose flower products (nectar and pollen) are the source of necessary nutrients for pollinating animals. However, broadly speaking, melliferous plants encompass certain gymnosperms as well, and apart from the floral they also offer non-floral products such as the extrafloral nectar, resin, honeydew and fruit sap (Figure 1). These, however, are not associated with pollination (Pacini and Nicolson, 2007; Mačukanović-Jocić, 2010). These plants are of crucial importance for the existence of many pollinating animals, and their disappearance would affect the survival of the honey bees, for which they have many levels of significance, e.g., in the development of their colony and the production of honey and other bee products. Honey bees are an essential part of the agroecosystems and completely depend on melliferous plants, which, by offering nectar and pollen as a reward for successful pollination, create conditions for a regular life, development and functioning of the bee society.

Photo: M. Mačukanović-Jocić and G. Jocić.

Figure 1. Floral and non-floral products of melliferous plants. a) Honey bee collecting nectar from *Prunus domestica* flowers; b) Honey bee seaking for extrafloral nectar on leaf stipules of *Vicia sativa*; c) Ants, connected with defence strategy of plant, collecting extrafloral nectar on leaves of *Prunus laurocerasus*; d) Solitary bee gathering pollen from plantain flowers; e) Honey bee searching for resin and honeydew from *Abies alba*; f) For lack of floral source of food, honey bee is feeding on grape fruit sap.

Therefore, melliferous plants are neither only those that produce nectar nor those that are visited exclusively by the honey bee. When the melliferous aspect is defined, one needs to take into account not only certain pollinating insects, but rather all animals whose survival depends on them, such as particular kinds of birds and mammal pollinators. Some plants give only pollen (poppy or primarily anemophilous plants – beech, oak, plantain), some give resin or honeydew (pine, fir), but are regarded as melliferous nonetheless (Ricciardelli D'Albore, 1997; Ricciardelli, D'Albore and Intoppa, 2000). Some entomophilous plants, on the other hand, have huge nectar production, even though, just like the ornithophilous plants or chiropterophilous plants, they are not visited by honey bees, and are intended for a different class or species of pollinators.

In the temperate continental climate, pollinating insects, including the honey bee, find food sources in wild forest, meadow, ruderal and swamp plants, and in agrophytocenoses, such as orchards, vineyards, fields of rapeseed, onobrychis, alfalfa and other agricultural crops, and in plantations of medicinal and aromatic plants. One option to improve bee pasture is the so-called *targeted growing* which would include single-purpose plants with the basic purpose of bee pasture (e.g., phacelia), or the so-called multi-purpose plants, primarily used for different purposes (remedial, forage crops and other cultures), but in the flowering phenophase these plants could also be used as melliferous. Great benefit can also be obtained from the so-called *additional growing*, which implies the sowing of honey plants in orchards, vegetable gardens, after the fruiting and the harvest of certain cultures (e.g., wheat), the planting of woody honey plants (black locust, false indigo bush, hazel, cornus...) on the unused land areas, selection of plants, i.e., varieties with a greater nectar production or longer period of flowering, etc.

The evaluation of the honey potential of plant species requires a very serious, complex and comprehensive research approach, since this is not a constant value which can be precisely expressed in numbers. It does not depend only on the nectar capacity and pollen production which reflect the endogenous properties of a plant. Rather, it is a variable parameter subject to the influence of external abiotic and biotic factors. There has been some speculation in literature about the melliferous aspects of a plant (or, from the point of view of beekeeping, what is the honey yield.

In order to find an answer, serious field and laboratory tests are required which would deal with the following parameters: time and duration of flowering, floral productivity, qualitative and quantitative characteristics of nectar production (total daily nectar amount per flower, type, amount and concentration of sugar in the nectar), the abundance of melliferous species, attractiveness factors, pollinator visits, etc.

Over the last two decades, as a contribution to the development and advancement of apiculture, a certain amount of research has been devoted to various aspects of studying the melliferous flora of Serbia and the wider region (Blaženčić et al., 1994; Mačukanović et al., 1996; Mačukanović and Grabeljšek, 1997; Grabeljšek, 1996; Perišić et al., 2004; Farkas and Zajácz, 2007; Grozeva, 2011; Jarić et al., 2013). The Serbian apiflora of forest, meadow, ruderal and anthropogenically changed phytocoenoses was studied with the aim of assessing their melliferous potential and their contribution to bee pasture during the flowering season. The melliferous plants were analyzed individually from the aspect of flowering phenology, their abundance in phytocoenoses, and the intensity of nectar and pollen production, as well as the production of honeydew.

The family Lamiaceae is cosmopolitan in distribution with two major centres of occurrence – Mediterranean and Central Asian regions, represented by c. 200 genera and 3200 species (Perveen and Qaiser, 2003). In Serbian flora it is represented by 32 genera and 162 species (Vasić, 1997). It is known as the predominantly entomophilous plant family, that has so far often been studied from the point of view of biology of melliferous plants and pollination, including the related scientific fields which are in contact or overlapping with them.

Floral construction and pollination biology in the Lamiaceae family were widely discussed (Brantjes and De Vos, 2006; Claßen-Bockhoff, 2007), including the fact that the morphological and physiological aspects of floral biology of some species favour cross-pollination (Da Silva at al., 2005). A few articles were focusing on the phenotypic diversity and adaptive significance of the bilabiate flowers, especially in *Salvia* which is characterized by staminal lever bio-mechanism separating it from other genera of the tribe Mentheae by only two fertile stamens (Claßen-Bockhoff et al., 2004; Rodriguez-Riaño and Dafni, 2007; Reith et al., 2007; Wester and Claßen-Bockhoff, 2007). Some of the attractiveness factors were analyzed separately and the combined effects of some were discussed (Kevan, 1983; Dafni et al., 1988; Arnold et al., 2009). Research devoted to the analysis of flower and nectar characteristics affecting bee attraction, as well as to the contribution of some melliferous species in bee pasture, were done in detail so far (Eisikowitch and Masad, 1980; Dafni et al., 1988; Danon et al., 1990; Farkas and Zajacz, 2007; Mačukanović-Jocić et al., 2011).

Nectar production has been studied in Lamiaceae species from different aspects so far, such as diurnal and seasonal dynamics with respect to different habitat conditions, especially under the influence of microclimatic parameters (Mačukanović-Jocić et al., 2004; Mačukanović-Jocić and Đurđević, 2005; Mačukanović-Jocić at al., 2008).

Morpho-anatomy of floral nectaries and ultrastructure of secretory cells were analyzed using light, fluorescence and scanning electron microscopy, in several species of Lamiaceae such as *Rosmarinus officinalis* (Zer and Fahn, 1992) and *Ocimum basilicum* (Mačukanović et al., 2007) aiming to provide better understanding of the mechanism of nectar secretion.

In the present study, pollen grains have been investigated under light and scanning electron microscope in order to provide a detailed morphological description contributing to the melissopalynological and taxonomic investigations. There are many studies devoted to different palynological aspects of the family Lamiaceae. In the mid-nineties, extensive light microscopic studies on Lamiaceae pollen were performed by Erdtman (1945, 1952). Besides Erdtman, the earliest palynological studies of this family have been carried out by Waterman (1960), Emboden (1965) and Soursa (1966). Many studies related to particular subfamilies, tribes, genera, species and subspecies provided important data that had been of use to subfamilial classification as well as to clear up evolutionary relationships (Cantino and Sanders, 1986; Abu-Asab and Cantino, 1989, 1992, 1993, 1994; Harley et al., 1992; Moon and Hong, 2003; Moon et al., 2008a, 2008b; Celenk et al., 2008: Salmaki et al., 2008). Some surveys, which rely on light microscopy and scanning electron microscopy, have been devoted to the species growing on the territory of different countries. For instance, the pollen morphology of 118 North American Lamiaceae has been described by Trudel and Morton (1992); in Pakistan, 79 species have been researched by Perveen and Qaiser (2003); 21 Mongolian taxa have been studied by Badamtsetseg et al., (2012), and plynological characteristics of *Salvia* taxa from Turkey were analyzed by Özler et al. (2011; 2013).

Pollen morphology has also been analyzed from melissopalynological standpoint (von der Oche and von der Oche, 2007; Bucher et al., 2004).

The present study was focused on the floral morphophysiological characteristics important in plant-pollinator interactions in five cultivated medicinal Lamiaceae species that could be simultaneously used as a potential source of food for honey bees. Selected bilabiate flowers are presented comparatively and their functional significance is discussed. The significance of floral attractive features affecting pollinator visitation as well as the degree of attractiveness has been established along with the fact which plants the honey bees prefer. Finally, there was an intention to establish, even though they belong to the same family, to what extent their difference in terms of basic floral attractive characteristics, reflects the degree of co-adaptation with co-evolutionary pollinating partners.

MATERIAL AND METHODS

Five cultivated species of the Lamiaceae family used to study the flower attractiveness are as follows: *Hyssopus officinalis* L., *Lavandula officinalis* Chaix, *Melissa officinalis* L., *Salvia officinalis* L. and *Salvia sclarea* L. Two-year long field and laboratory investigations were conducted in the period between 2004 and 2005. The field experiment including studies on the nectar production parameters, observation of the flower visitors and frequency of honey bee visits were conducted during the flowering period in 2004 at the Institute for Medicinal Plant Research "Dr Josif Pančić" in the vicinity of Belgrade (Serbia), where selected medicinal and aromatic plants were grown in plantation. During the second year of research, morphoanatomical examination of nectaries was made, as well as the analysis of morphometric characteristics of flowers and palynomorphological study.

Photo: M. Mačukanović-Jocić and G. Jocić.

Figure 2. *Hysoppus officinalis* L. (hyssop).

Biological and Ecological Characteristics of Study Species

Hysoppus officinalis L. (hyssop) is an aromatic perennial semi-evergreen sub-shrub, 20-80 cm tall. Its dark blue flowers are arranged in whorls, in the axils of upper leaves, forming together thick, spike-like inflorescence on the peak branches located on one of its sides (Figure 2). It blossoms between June and October (Jančić et al., 1995). In Serbia, it occurs subspontaneously at several localities (Jančić et al., 1995), participating in the plant communities *Artemisio-Salvietum officinalis*, or more rarely *Carpinetum orientalis serbicum* (Diklić, 1974). As a locally rare species, it is protected, and has been assessed (CITES category) as vulnerable – VU (Dajić et al., 2003). It grows on arid, very sunny, mostly limestone rocks, and on depleted pastures.

Lavandula officinalis Chaix. (common lavender) is a perennial woody based sub-shrub, 40-60 cm tall. Its flowers are arranged in 3-4 mutually distant whorls forming interrupted spikes at the top of the plant and side branches. Its corolla is tubular, with underdeveloped lips, of blue-violet colour, covered in hair (Figure 3). The highest percentage of the flower essential oil can be found in the sepal glands (Kišgeci, 2002). It blossoms between June and September (Diklić, 1974). It grows in rocky ground, dry, hot and craggy slopes up to 1700 m altitude (Tucakov, 1997; Kišgeci, 2002). In Serbia it does not occur in the wild, but it is grown on plantations for its essential oil obtained by the distillation of fresh flowers (*Lavandulae flos*) (Kišgeci, 2002).

Mellissa officinalis L. (lemon balm) is a perennial herbaceous herb, 30-80 (120) cm tall, with a vigorous, lemon scent. Its inconspicuous white or pale yellow flowers appear in the axils of much longer upper leaves (Figure 4).

Photo: M. Mačukanović-Jocić and G. Jocić.

Figure 3. *Lavandula officinalis* Chaix. (common lavender).

Photo: M. Mačukanović-Jocić and G. Jocić.

Figure 4. *Mellissa officinalis* L. (lemon balm).

It flowers between June and August (Jančić, 1995). It grows on roadsides, in vineyards, ruderal habitats, next to settlements, along with fences and hedges. In Serbia, it is distributed only sporadically, from the plains to the altitude of above 1000 m (Tucakov, 1997). It is cultivated as a remedial and aromatic plant, for its essential oil of sedative and inflammatory properties obtained by the distillation of leaves (*Melissae folium*) (Kišgeci, 2002).

Since long ago it has been used as an important remedial, ornamental and melliferous plant, and it was believed that it should be grown close to bee hives, which, after being rubbed with lemon balm leaves keep bees together.

Salvia officinalis L. (common sage) is a perennial semi-shrub with a much branched stem, with the height of 50-90 cm. Its flowers are thinly distributed in whorls and form spike-like inflorescence on the tops of stems and side branches. The corolla is dominantly two-lipped (Figure 5), blue-violet or sometimes pink (Tucakov, 1997).

The whole plant has a very aromatic and recognizable odour. It is considered as a highly appreciated melliferous plant that provides quality unifloral honey. It blossoms between May and July (Jančić, 1995). Common sage usually grows on sunny limestone rocky grounds, poor in fertile soil, resisting long-lasting summer draughts and heat.

In Serbia, in the wild, it grows in two localities which might be considered refugial, as one of edificatory species in a polydominant association *Artemisio-Salvietum officinalis* (Diklić, 1974; Dajić and Kojić, 1999). This species has been protected by law in the locally rare CITES category – LR (Dajić et al., 2003).

Photo: M. Mačukanović-Jocić and G. Jocić.

Figure 5. *Salvia officinalis* L. (common sage).

It is often grown in gardens as a remedial and decorative plant and in plantations for its essential oil obtained by distillation from its dried herbaceous upper parts (*Salviae herba*) or leaves (*Salviae folium*) (Jančić, 1995).

Salvia sclarea L. (muscatel sage; clary sage) is a herbaceous hairy biannual strongly aromatic plant, up to 120 cm of height. Its bicoloured, cream and lilac to pink flowers are arranged in whorls, in axils of coloured (lilac to white) membranous bracts (Figure 24), building on the top of the stem a long upright richly branched terminal panicle thickly covered in glandular trichomes (Jančić, 1995).

It blossoms between June and August, and in favourable conditions a second blossoming is possible in September (Kišgeci, 2002). The whole plant has an unpleasantly overpowering scent (Houdret, 2002). It is found on dry, sandy, infertile sunny habitats, in rockies and on the roadsides. In Serbia, it is found on arid, rocky slopes of certain mountains in the eastern and south-eastern parts of the country.

Morphometric Properties of Flowers and Inflorescences

Flower features were morphologically investigated and measured morphometrically. Morphometric measurements of flowers and inflorescences, carried out on fresh plant material, included (Figure 6): flower length (length of the corolla) (a), calyx length (b), corolla tube length (c), corolla height (maximum distance between the upper and lower corolla lips) (d), width of the lower corolla lip (landing platform) (e), diameter of the corolla tube opening (f) and inflorescence length (g). Corolla colour, flowering time, flower lifespan, number of whorls per inflorescence and total number of flowers per plant have also been established. Randomly chosen flowers were marked at the bud stage and observed up to the first senescence signs. Data on qualitative and quantitative composition of the essential oil isolated from different plant parts were taken from literature.

Morphoanatomical Characteristics of Floral Nectaries

Morphoanatomical characteristics of floral nectaries, which encompass the defining of their position in the flower, and the analysis of size, shape, surface structure and glandular tissue structure, have been carried out by light, fluorescence and scanning electron microscopy as well as by stereomicroscopy in the previously fixed and fresh material. The plants from which flowers were taken for nectary analysis were at the full blossoming stage (two days after the bud opening). For LM study, flowers were prepared by removing the floral parts near the gland (the apical portions of the sepals, petals, stamens).

Photo: M. Mačukanović-Jocić and G. Jocić.

Figure 6. Morphometric measurements of flower and inflorescence.

The remaining receptacles with the ovaries were fixed in FAA for 24 h and post-fixed in 70% ethanol, dehydrated through a graded series of ethanol and then xylol before embedding in paraffin wax and dissected to various degrees. Longitudinal sections (3–5 μm thick) were made on LEICA SM 2000 R microtome, mounted serially and stained with alcian blue-safranin.

The samples were examined by a Leica DMSL microscope equipped with a digital camera (Leica DC 300) and Leica IM1000 software.

To examine the cuticle and xylem, freshly sectioned material was monitored for autofluorescence using a LEICA DMLS epifluorescence microscope (filter I3 BLU 450–490 nm). In order to define the position in the flower and to analyze the size, shape and surface structure of nectaries, the receptacle with ovary has been photographed and studied on fresh material without prior preparation procedure, using a Nikon Research Stereo Microscope SMZ 18 and JEOL JSM-35 scanning electron microscope.

Palynomorphological Analysis

Pollen grains of five species have been examined by light microscopy (LM) and scanning electron microscopy (SEM). The following features describing pollen grains were analyzed: symmetry, polarity, size, shape, type of ornamentation and aperturation. Polar axis (P), equatorial axis (E), colpus length, (CL), exine thickness (EGT), mesocolpium diameter (ME), and diameter of lumina (LD) were measured and number of lumina per sample area (25 μm^2) were counted. The terminology used is in accordance with Erdtman (1971), Faegri and Iversen (1989) and Punt et al. (2007).

For LM study, pollen grains from mature anthers were mounted in Kaiser's glycerin jelly. The exine thickness was measured under LM. For SEM study, the samples were mounted directly on metallic stubs using double sided adhesive tape and coated with gold (in BAL-TEC SCD 005 Sputter Coater, 100 seconds in 30 mA) in a sputtering chamber, and examined using JEOL JSM-6390 LV electron microscope at an acceleration voltage of 20 kV. SEM micrographs were used mainly for studying the general shape and size of grains, aperture characteristics, as well as to get more detailed information on the ornamentation.

Observations and measurements were done on a sample of 30 or more grains for each morphological character, and in addition, arithmetic means and standard deviations were calculated.

Nectar Production

Examination of nectar production included the analysis of the following parameters: total daily nectar amount (24-hour nectar accumulation), summary nectar amount per flower secreted during daytime (sum of periodical measurements performed during the daily bee activity between 7 am and 7 pm), nectar secretion rate, secretion intensity during flower ontogeny and diurnal and seasonal dynamics of nectar secretion. Nectar was collected using the microcapillary method described by McKenna and Thomson (1988). Total daily nectar amount has been defined by the sampling of nectar in the morning, after the plants have been covered for 24 hours.

In order to analyze diurnal secretion dynamics, nectar was collected at three-hour intervals during daytime (from 7 am to 7 pm) every 15 days. Inflorescences were covered with fine mesh for 12 hours prior to nectar removal (from the evening of the previous day) and between daily measurements, to prevent access by flower visitors, i.e., to protect flowers against nectar-collecting insects.

Seasonal dynamics of nectar secretion has been defined by measuring the total daily nectar amount per flower, every seven days during the entire flowering phenophase of plants studied. The secretion intensity during flower ontogeny was established by the measuring of the amount of nectar per flower, once a day, during its lifetime, from the day the flower opens until the day when the first signs of wilting appear.

The nectar secretion rate (μl/h) has been calculated by dividing the summary diurnal nectar amount per flower with the time interval within which nectar was periodically sampled (12 hours). Nectar sugar concentration was measured immediately upon its removal using portable hand refractometer.

Microclimatic Measurements

This research was performed in the temperate continental climate. At the time of nectar collecting and pollinator observation (three-hour daily intervals), every 15 days in the flowering season, the microclimatic parameters have also been monitored: air temperature (thermometer) and air humidity (hygrometer).

Pollinator Observation

Insects seen visiting the flowers were observed and counted every 15 days, at three-hour intervals during daytime at the time of nectar sampling, usually repeated three times in quick succession to give a mean insect count per plant at each sampling time. Insect-pollinators were identified without capture. Flower visitors and frequency of their visits (in 10 min/plant) were registered and documented by camera. The percentage of flowers visited by honey bees (in 10 min/plant), on three randomly chosen plant individuals of each species, was estimated with respect to the total number of open flowers per plant, as suggested by Dafni et al. (1988).

RESULTS AND DISCUSSION

Morphometric Characteristics of Flowers and Inflorescence

Flowers of five species have variously united petals forming 2-lipped corollas. Flower morphology characteristics related to bilabiate corolla and nectar positioned at the base of tubular part are attributive to this family (Fahn, 1988). Generally, the upper lip has two lobes and forms a hood over the lower lip providing a landing platform for pollinating insects. The upper lip is more developed, in comparison with the lower one, in *S. sclarea* in which it is highly falciform and *L. officinalis* in which it is not very prominent, whereas in the other three species it is the other way round. While in most species corolla tube is straight, in the case of lemon balm and to a lesser extent in lavender, it is arched, ending in an infundibular extended opening pointing at the top (Figure 8). Stamens (2 or 4, epipetalous) and pistil are mostly overlaid except in *H. officinalis* which has prominent and out-of the flower fertile parts (Figure 7).

Comparative analysis of flower morphometric characteristics revealed that the largest flower belongs to *S. sclarea*, followed by: *S. officinalis*, *M. officinalis*, *H. officinalis* and *L. officinalis* (Table 1; Figure 8).

The longest inflorescence, made of flowers appearing in several clusters of 6 on the stem, is in *S. sclarea* (38.1 cm), and the shortest in *L. officinalis* (7.2 cm). The highest values of flower length, calyx length, and corolla height were also recorded in *S. sclarea*.

Taking into consideration the average number of flowers per whorl and per plant at fool flowering stage, flower lifespan and the duration of flowering period, hyssop is characterized by the highest production of flowers, providing the greatest benefit to the bee and other pollinating insects among the researched species.

Regarding the flower colour, in the most species there is great variability. In hyssop, three varieties of blue, pink and sporadically white corollas coexist. Corolla of lemon balm is creamy to white, of clary sage white to pale lilac or pink, of lavender blue-lilac and of sage different nuances of blue.

Unlike the strong and heavy fragrance of *S. sclarea*, the whole plants of *S. officinalis*, *L. officinalis*, *H. officinalis* and crushed leaves of *M. officinalis* appear to be pleasantly scented.

The longest flower lifespan was recorded in *L. officinalis* (average 5.8 days), which is almost equal to hyssop (5.6 days) and sage (5.2 days), and the shortest for *M. officinalis* (not more than 20 hours). Flowers of *S. sclarea* lasted 4 days on average. Flower lifespan was shortened if in the meantime pollination was done. With regard to flowering dynamics, *S. officinalis* was the first plant began to flower in the second half of May (Figure 9).

Other plants started to blossom about 2 weeks later, in June, by order of succession: *H. officinalis, L. officinalis, S. sclarea* and *M. officinalis*, reaching in full bloom after fifteen days. Flowering period of all species generally takes from 2 to 5 months, but in the favourable weather conditions, their blossoming continuously or discontinuously extends until the beginning of autumn or even later. The longest flowering period was observed in *L. officinalis.*

Photo: M. Mačukanović-Jocić and G. Jocić.

Figure 7. Flowers of studied species arranged in whorls: a) *H. officinalis*, b) *L. officinalis*, c) *M. officinalis*, d) *S. officinalis*, e) *S. sclarea*.

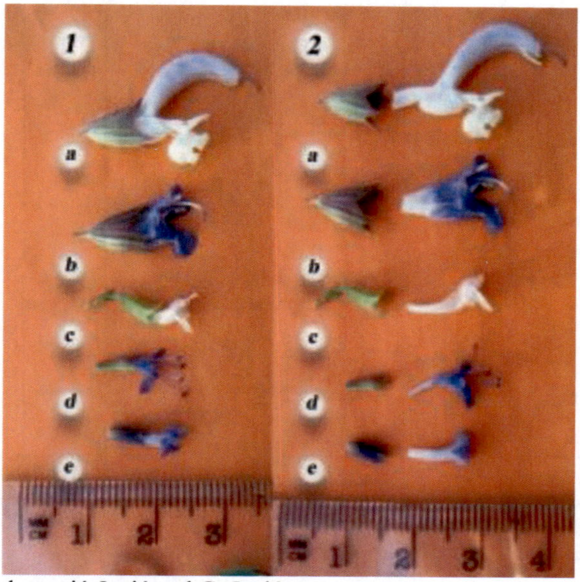

Photo: M. Mačukanović-Jocić and G. Jocić.

Figure 8. Shape and size of flowers comparatively shown: a) *S. sclarea*, b) *S. officinalis*, c) *M. officinalis*, d) *H. officinalis*, e) *L. officinalis* (1 - complete flowers; 2 - calyx and corolla separately).

Table 1. Morphometric characteristics of flower and inflorescence of five Lamiaceae species

	H. officinalis	*L. officinalis*	*M. officinalis*	*S. officinalis*	*S. sclarea*
Flower length (mm)	**12.50** ± 0.79	**11.20** ± 0.91	**15.40** ± 0.42	**21.30** ± 1.57	**24.10** ± 0.65
Calyx length (mm)	**6.40** ± 0.65	**4.74** ± 0.50	**8.50** ± 0.35	**11.44** ± 0.26	**14.08** ± 0.43
Corolla tube length (mm)	**6.76** ± 0.51	**7.50** ± 0.61	**10.50** ± 0.5	**11.00** ± 0.35	**9.00** ± 0.14
Width of lower corolla lip (mm)	**5.80** ± 1.44	**7.60** ± 0.65	**7.66** ± 0.76	**9.30** ± 0.44	**8.08** ± 0.24
Corolla height (mm)	**7.96** ± 1.13	**8.50** ± 0.93	**7.30** ± 0.27	**15.00** ± 1.17	**20.00** ± 0.61
Diameter of corolla tube opening (mm)	**2.02** ± 0.47	**1.68** ± 0.25	**2.00** ± 0.35	**3.70** ± 0.67	**2.16** ± 0.23
Inflorescence length (mm)	**203.20** ± 76.24	**72.00** ± 23.61	**357** ± 134.59	**199** ± 53.82	**381** ± 122.83
Whorls per inflorescence	**19.20** ± 1.10	**5.60** ± 0.56	**10.8** ± 0.45	**7.80** ± 2.28	**7.40** ± 0.89
Total number of flowers per whorl	**28.40** ± 2.30	**11.31** ± 2.98	**26.00** ± 1.40	**10.80** ± 1.79	**6.00** ± 0.00
Number of open flowers per whorl in nectar collecting days	**3.60** ± 1.14	**3.00** ± 1.00	**2.40** ± 1.14	**2.60** ± 0.89	**3.20** ± 1.92

Mean values ± SD are shown.

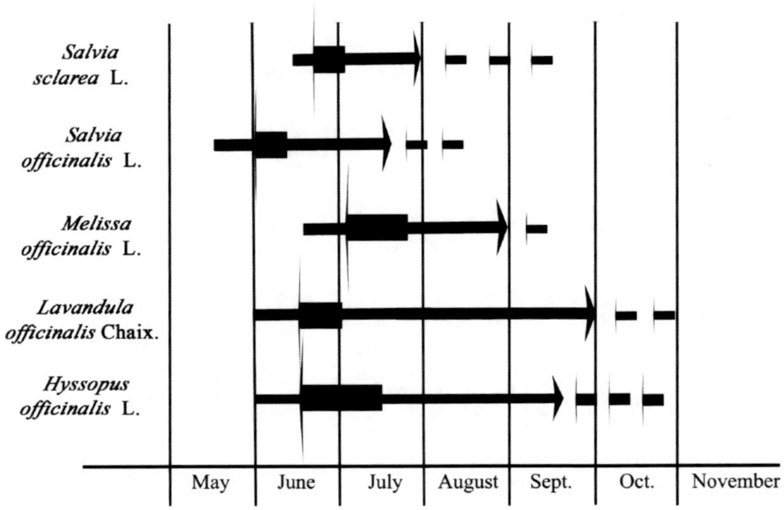

Figure 9. Flowering phenology of study species in 2004.

Palynomorphological Analysis

LM and SEM micrographs of the examined pollen grains are shown in Figures 10 and 11. The main pollen characteristics are given in Table 2. High values of standard deviation among certain parameters demonstrate a high variability of tested characteristics.

The pollen grains are monads, radially symmetrical and isopolar, varying in size from medium (hyssop, lavender and lemon balm) to large (sage and clary sage). The largest pollen grain, on average, was observed in *S. sclarea* while *L. officinalis* has the smallest one.

The mean of polar axis (P) ranges from 28.38 μm in *L. officinalis* to 58.32 μm in *S. sclarea*, and the size of the equatorial axis (E) ranges in average from 26.39 μm in (*L. officinalis*) to 41.77 μm (*S. sclarea*) (Table 2). In *S. sclarea* there are 2 evidently different size classes in the same pollen assemblage, of similar shape, dominated by Class I, which encompasses large grains (mean P-58.32 μm), while the less frequent Class II encompasses usually the deformed, medium-sized grains (mean P – 34.5 μm). In *L. officinalis*, also, two types of pollen grains (type a, and type b) coexist in the same specimen, which, in addition to size, differ significantly in terms of shape (the following passage). In *S. officinalis,* due to a high range of variation of P and E sizes, and the P/E index, the most pronounced is the intraspecific morphological heterogeneity.

SEM examination revealed that often pollen shape in equatorial view varies considerably within the same species, even in the same specimen, coexisting within the range from sub-oblate to prolate (P/E ranged from 0.76 – 1.90). For the species studied, the most frequent shapes in equatorial view is found to be as follows: sub-prolate in *H. officinalis* and *S. sclarea* (Class II); prolate in *L. officinalis* (type a), *M. officinalis, S. officinalis* and *S. sclarea* (Class I), and oblate-spheroidal in *L. officinalis* (type b); (Table 2). While in *L. officinalis* in the same specimen, the two extremely different types are almost equally present, the most different pollen shapes have been encountered in *S. officinalis* coexisting between spheroidal and prolate (P/E ranged from 0.95 to 1.9) (Figure 10-4a). The shape in polar view is rarely circular or subcircular to prominently oval among those that are very flattened laterally such as in prolate polen grains of *H. officinalis, S. officinalis* and *S. sclarea* (Figure 10 - 1c, 3c-5c and Figure 11 - 1b, 3b-5b). In *L. officinalis* equatorial outline is found to be clearly hexagonal (Figures 10 -2c and 11 - 2b).

The pollen grains are mostly hexacolpate in all species except in *S. sclarea* in which they are very rarely intermixed with tetracolpate and octacolpate ones (Figure 10 - a,b).

The ectocolpi are distributed symmetrically, elongated, usually shallow, and narrowing at the poles. In *S. sclarea*, the colpal membrane is finely or coarsely granular, visible in SEM (Figure 10-5c). The mean colpus length of all the studied species, being correlated with the length of the polar axis, varies from average 19.95 μm (*L. officinalis* type b) to 46.91 μm (*S. officinalis*) (Table 2). The colpus width is the most variable in *L. officinalis* ranging in average from 0.72 μm to 6.26 μm. In the polar view, all 6 intercolpal areas are the same width (*L. officinalis*), or, if the grains are narrowing more at the equator, 2 of the mesocolpia are larger than the other 4, like in prolate 6-colpate grains of *S. officinalis* and *S. sclarea* (Figures 10 - c). The mesocolpial area varies in average from 9.78 μm in *H. officinalis* to 12.86 μm in *S. sclarea,* being slightly broadened at the equator.

The exine sculpturing displays several types of surface structures: bireticulate (*H. officinalis, M. officinalis, S. officinalis*), reticulate to foveolate (*S. sclarea*) and foveolate (to perforate) (*L. officinalis*). (Figure 10 - 1d-5d). The bireticulate pattern is characterized by the primary lumen of closed angular shape with well differentiated secondary reticulum in it. The largest primary lumina as well as secondary ones, in average, were found in *S. officinalis* (1.83 μm; 0.58 μm). The thickest muri in average were present in *L. officinalis* (1.2 μm), and the thinnest primary muri were recorded in *H. officinalis* (0.14 μm).

Number of primary lumina (or perforations) per sample area of 5 μm x 5 μm ranges in average, from 6 (*L. officinalis*) to 15.25 (*H. officinalis*) (Table 2). Diameter of secondary lumina in average varies from 0.29 μm (*H. officinalis*), to 0.53 μm (*M. officinalis*) and 0.58 μm (*S. officinalis*). The greatest number of secondary lumina per primary lumen is recorded in *H. officinalis.* Size of the lumina decreases towards the apertures and poles in all species, except in *L. officinalis* and *M. officinalis*. For decreasing tendency also in lumina number, in *S. sclarea* at the apocolpium, a transition from reticulate to foveolate (sometimes psilate) ornamentation can be perceived (Figure 10-5c). Variation in the shape of lumina is a characteristic feature for five species, such that it changes from polygonal, to elongated or rounded. Lumina are polygonal in shape and rather uniform in *H. officinalis, M. officinalis* and *S. officinalis*; however they shows variations as being round or round-ellipsoidal and elongated in the mesocolpial and apocolpial area of *L. officinalis* and *S. sclarea* (Figure 10 - 1d, 2d, 3d, 4d, 5c and 5d). Similarly, muri are thick and prominent in the last two species. In *L. officinalis*, lumina are distributed regularly over the whole surface, but besides shallow depressions, also deep perforations (<1 μm) occur –with the former being more frequent (Figure 10-2d).

Exine thickness ranged in average from 0.94 μm in *H. officinalis* to 1.75 μm in *L. officinalis* (Figure 11 - 2b, 2c). The exine is somewhat thicker in the polar than in the equatorial region.

According to Perveen and Qaiser (2003), Lamiaceae is a eurypalynos family, although some genera, such as *Lycopus*, are stated to be stenopalynos (Moon and Hong, 2003) indicating a limited taxonomic application of pollen characters due to their slight variation at interspecific level.

Pollen of five species in the present study share some common morphological features with the other Lamiaceae pollen (Perveen and Qaiser, 2003; Moon et al., 2008b; Özler et al., 2013) being generally isopolar and radially symmetrical, and mostly hexacolpate. Besides uniform general morphological characteristics including type and number of apertures, fine details of exine sculpturing and colpal features are characteristic to differentiate the taxa from each other (Özler et al., 2013). As already concluded in earlier studies (Gupta and Sharma, 1990; Perveen and Qaiser, 2003; Moon et al., 2008; Badamtsetseg et al., 2012), pollen showed considerable variation within the family, also at species level, or, even, in the single specimen, both in terms of size, (very small, small, medium to large), and shape (oblate, sub-oblate, oblate-spheroidal, prolate-spheroidal, sub-prolate to prolate), and also in terms of exine ornamentation (reticulate, rugulate, foveolate, perforate, striate, fosulate) or aperture number.

By describing palynomorphological features in the species from Pakistan, Perveen and Qaiser (2003) have also demonstrated significant variability within the family. On the basis of shape, type and number of apertures and ornamentation, characteristics which according to these authors have taxonomic value, Lamiaceae species are classified into 7 distinct pollen types. According to this classification, *Melissa officinalis*, belonging to the *Salvia aegyptiaca* type, is characterized by 6-colpate, oblate-spheroidal to sub-oblate and reticulate pollen. This description concerning shape is not in accordance with what is presented here, since *Melissa* pollen is found to be sub-prolate to prolate.

According to Perveen and Qaiser (2003) Lamiaceae pollen is mostly colpate (3 or 6 colpi) and rarely colporate. However, although it was considered that the number of colpi is the significant character of pollen morphology to solve taxonomic problems of identification and classification, recent studies have revealed that this character is probably of no great importance, since it is not reliable distinguishing parameter at the species level exibiting great intraspecific variability (Gupta and Sharma, 1990).

The present study shows that the pollen grains of all species are predominantly hexacolpate, but in the case of *S. sclarea* they are mixed together with tetracolpate and octacolpate ones. All 7 *Salvia* species, classified under the group designated as *Salvia aegyptiaca* type by Perven and Qaiser (2003) were found to be hexacolpate. However, a variable number of colpi within the same *Salvia* species were recorded by Őzler et al. (2011; 2013) who found, in addition to hexacolpate, coexisting heptacolpate grains in *S. palaestina,* and octacolpate in *S. recognita* from sect *Salvia* and also those with seven and eight colpi in *S. candidisima* subsp. *candidisima*. Also, Moon et al. (2008a) found tetracolpate (S. *eremostachya, S leucantha* and *S palaestina*), pentacolpate (*S. eremostachya* and *S. splendes*), and octacolpate (*S. barrelieri, S. palaestina* and *S. uliginosa*) grains. Similarly, study of the pollen from single specimen of *Salvia leucantha* performed by Gupta and Sharma (1990), have shown significant variation in aperture number, so that besides 6-colpate, 4-, 5-, 7-, 8-, 9-, 10-, 11- colpate to spiraperturate grains also occurred. The present study reveals a range of morphological variations even in single pollen assemblage. The pollen grains of the species examined are medium to large, but variations are also observed within the species, such as in the case of *S. officinalis* and *S. sclarea*.

As pointed out elsewhere (Moon and Hong, 2003), the shape of pollen in equatorial view is variable among the taxa, even within the same taxon. While pollen grains of *S. officinalis* vary considerably both in size and shape, in *L. officinalis* two shapes are clearly distinguished. Although different species were investigated, similar intraspecific heterogeneity for pollen shape within the genus *Lavandula* was described by León-Arencibia and La-Serna Ramos (1992). Also, Gupta and Sharma (1990) pointed out that pollen shape varied within *Salvia leucantha* from oblate, suboblate, oblate-spheroidal, prolate-spheroidal to subprolate. According to Perveen and Qaiser (2003) shape of pollen grains of 79 analyzed Lamiaceae species are prolate-spheroidal to oblate-spheroidal or sub-oblate. Őzler et al. (2013) palynomorphologically investigated 30 Salvia taxa in sections *Hymenosphace* and *Aethiops* grown in Turkey and found in most taxa suboblate, oblate-spheroidal, or prolate-spheroidal to spheroidal grains and occasionally subprolate.

Pollen grains of the majority of species have been found to fall into a general reticulate ornamentation type, while those of *L. officinalis* have a foveolate (to perforate) ornamentation. Although exine ornamentation appear to have significant taxonomic value in segregation of some taxa (Moon et al., 2008), it is often regarded as insufficient criteria for separating the species as being the case in stenopalynous genus *Lycopus* (Moon and Hong, 2003). As was already concluded in some earlier studies (Trudel and Morton, 1992) exine surface is very homogenous and reticulate tectum is the most common type among Lamiaceae, but some other types of ornamentation have also been found: rugulate, foveolate, striate, fossulate and perforate (Perveen and Quiser, 2003; Moon et al., 2008b). Őzler et al. (2013), who palynomorphologically investigated *Salvia* species, found two patterns of exine sculpturing: the bireticulate and the reticulate-perforate, both divided into subtypes according to the number of primary lumina per sample area. According to these authors, *S. sclarea* falls into the first type, which is not entirely in agreement with present study results demonstrating that this species tectum is mainly reticulate (to foveolate towards the poles). Also, *S. officinalis* would, according to the given division, be classified as reticulate-perforate type among which these authors classified 12 *Salvia* taxa, which is characterized by a 2-layered tectum consisting of a suprareticulate intratectal layer and pores in an infratectal layer restricted next to muri. Ornamentation description of seven studied endemic taxa of *Lavandula* L. described by Leon-Arencibia and La-Serna Ramos (1992) does not coincide entirely with our results regarding *L. officinalis*, since tectum is found to be foveolate rather than perforate.

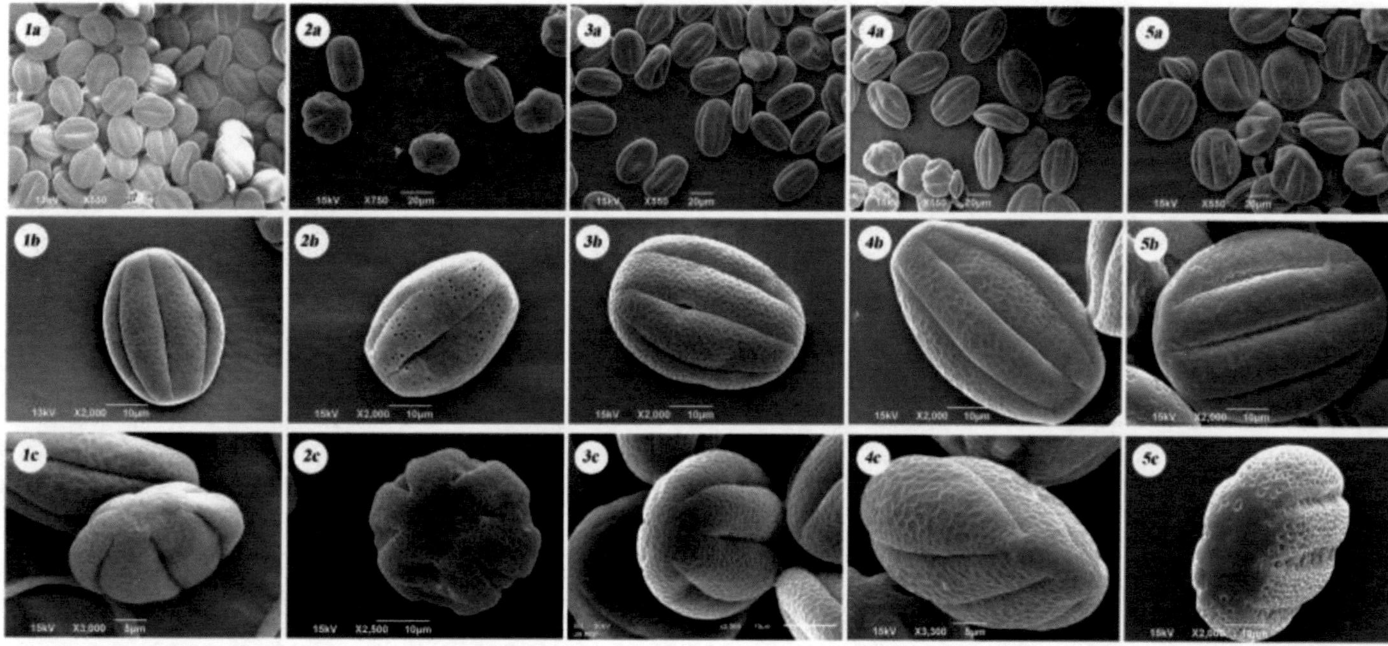

Figure 10. (Continued)

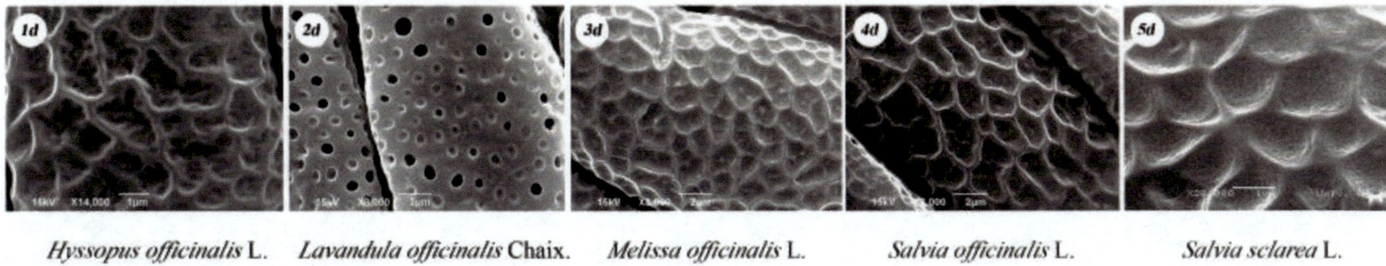

Figure 10. Scanning electron photomicrographs of pollen grains of five Labiatae. 1. *Hyssopus officinalis* (x 550); 2. *Lavandula officinalis* (x 750); 3. *Melissa officinalis* (x 550); 4. *Salvia officinalis* (x 550); 5. *Salvia sclarea* (x 550). 1a-5a: Pollen assemblage of species examined. Shape and size variation is noticeable in *L. officinalis* (2a), *S. officinalis* (4a) and *S. sclarea* (5a). In *S. sclarea*, also variation in aperture number is visible. 1b-5b: 6-colpate pollen grains in equatorial view showing most frequent shapes: sub-prolate (*H. officinalis, S. sclarea*), oblate spheroidal (*L. officinalis* type b), prolate (*L. officinalis* type a, *M. officinalis, S. officinalis*). Colpi are narrow towards poles. 1c-5c: colpate pollen grains in polar view showing number and position of apertures. 1d-5d. Detail of exine surface showing bireticulate (*H. officinalis, M. officinalis, S. officinalis*), foveolate (to perforate)(*L. officinalis*) and reticulate (*S. sclarea*) ornamentation.

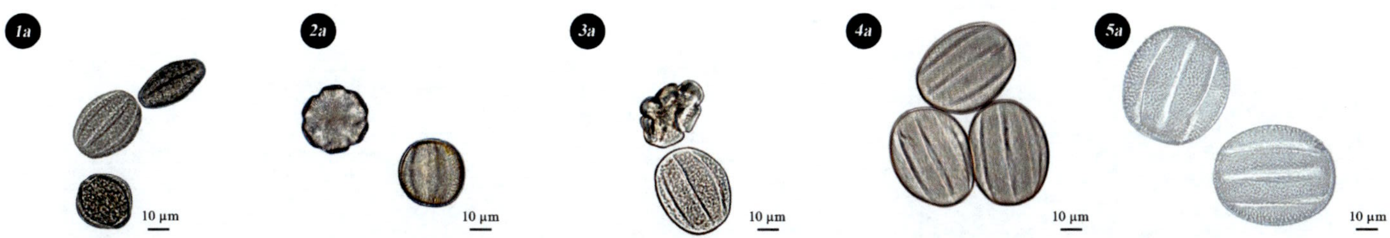

Figure 11. (Continued)

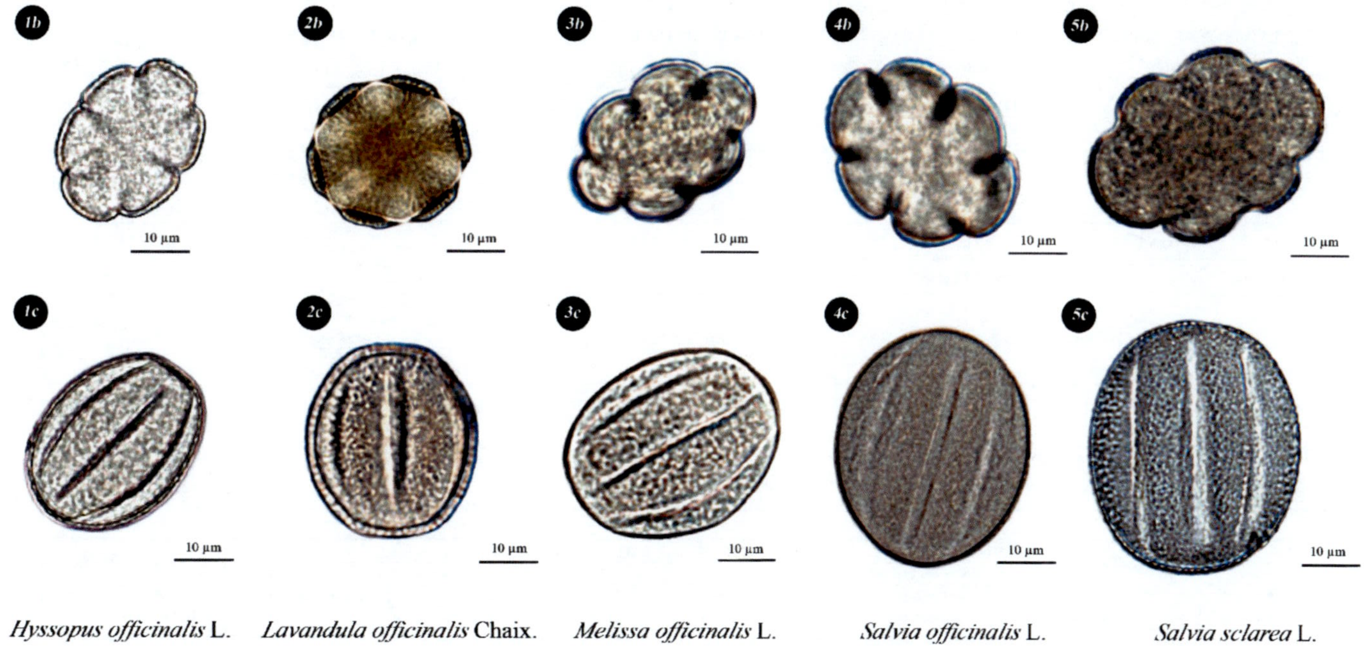

Figure 11. Light photomicrographs of pollen grains of five Labiatae. 1a-c: *Hyssopus officinalis*; 2a-c: *Lavandula officinalis*; 3a-c: *Melissa officinalis*; 4a-c: *Salvia officinalis*; 5a-c: *Salvia sclarea*; 1b-5b: polar view showing oval (*H. officinalis*, *M. officinalis, S. officinalis, S. sclarea*) and hexagonal (*L. officinalis*) outline. 1c-5c: equatorial view. Optical section.

Table 2. Pollen morphological data of the studied species

Species \ Morfometric data			P (µm) min (Ō ± SD) max	E (µm) min (Ō ± SD) max	CL (µm) min (Ō ± SD) max	C W min (Ō ± SD) max	EGT (µm) min (Ō ± SD) max	P/E min (Ō ± SD) max
Hyssopus officinalis			34.10 (**36.10** ± 1.15) 38.20	24.00 (**28.00** ± 1.73) 30.10	26.30 (**30.08** ± 2.00) 35.00	0.44 (**0.8** ± 0.5) 1.80	0.73 (**0.94** ± 0.18) 1.32	1.11 (**1.29** ± 0.09) 1.56
Lavandula officinalis	Type	a	34.15 (**36.88** ± 1.32) 39.05	19.80 (**26.39** ± 3.76) 29.60	27.94 (**30.20** ± 1.57) 33.00	0.20 (**0.72** ± 0.33) 1.08	1.31 (**1.75** ± 0.25) 2.20	1.31 (**1.40** ± 0.20) 1.72
		b	25.10 (**28.38** ± 2.30) 31.60	26.20 (**29.88** ± 3.51) 34.60	17.00 (**19.95**± 2.13) 21.80	5.80 (**6.26** ± 0.48) 7.40		0.76 (**0.97** ± 0.18) 1.20
Melissa officinalis			40.21 (**46.48** ± 2.18) 51.02	29.29 (**32.83** ± 2.11) 37.20	36.77 (**40.68** ± 2.17) 44.03	0.40 (**0.90** ± 0.33) 1.60	0.83 (**1.08** ± 0.18) 1.66	1.20 (**1.41** ± 0.09) 1.60
Salvia officinalis			34.60 (**50.77** ± 8.9) 61.00	29.50 (**34.65** ± 2.59) 39.00	33.45 (**46.91** ± 4.88) 55.70	0.70 (**1.34** ± 0.45) 2.10	0.83 (**1.06** ± 0.16) 1.37	0.95 (**1.47** ± 0.28) 1.90
Salvia sclarea	Class	I	54.00 (**58.32** ± 2.3) 62.48	31.00 (**41.77** ± 5.47) 50.00	36.00 (**44.77** ± 4.54) 52.00	0.82 (**2.64** ± 1.82) 6.92	0.83 (**1.34** ± 0.29) 1.96	1.16 (**1.41** ± 0.18) 1.87
		II	32.00 (**34.5** ± 2.75) 41.00	19.00 (**27.10** ± 4.20) 33.00	22.50 (**26.22** ± 1.92) 28.30	2.10 (**3.25** ± 0.74) 4.20		1.03 (**1.29** ± 0.20) 1.74

Species \ Morfometric data			shape	Me min (Ō ± SD) max	PLD min (Ō ± SD) max	Mu min (Ō ± SD) max	No Pl /25 µm^2 min (Ō ± SD) max	SLD µm min (Ō ± SD) max	NoSL/PL min (Ō ± SD) max
Hyssopus officinalis			(prolate-sferoidal) **sub-prolate** (prolate)	8.05 (**9.78** ± 0.85) 11.12	0.80 (**1.58** ± 0.63)3.35	0.10(**0.14** ± 0.03) 0.22	11.00(**15.25** ± 2.3)19.00	0.19(**0.29** ± 0.07)0.41	6.00(**11.54** ± 2.76)18.00
Lavandula officinalis	Type	a	(sub-prolate) **prolate**	7.50 (**10.05** ± 1.21) 11.81	0.39 (**0.63** ± 0.11)0.90	0.75(**1.20** ± 0.35)1.90	5.00(**6.00** ± 1.24) 8.00		
		b	(sub-oblate) **oblate-spheroidal** (sub-prolate)	8.50 (**9.92** ± 0.67) 10.50					
Melissa officinalis			(sub-prolate) **prolate**	10.00 (**10.82** ± 0.85)12.34	0.66(**1.39** ± 0.58)2.87	0.15(**0.24** ± 0.06)0.34	8.00(**12.62** ± 2.66)16.00	0.32(**0.53** ± 0.27)1.32	3.00(**4.00** ± 0.81)5.00
Salvia officinalis			(sferoidal) **prolate**	7.79(**10.27** ± 1.76)14.01	1.50(**1.83** ± 0.48)3.45	0.24(**0.29** ± 0.04)0.40	5.00(**6.20** ± 0.91)7.00	0.33(**0.58** ± 0.16)0.95	5.00(**7.17** ± 2.06)11.00
Salvia sclarea	Class	I	(sub-prolate) to **prolate**	10.70 (**12.86** ± 1.49)16.35	0.80 (**1.02** ± 0.11)1.34	0.14(**0.37** ± 0.13)0.65	4.00(**7.48** ± 2.27)12.00		
		II	(prolate-sferoidal) **sub-prolate**	8.60 (**9.80** ± 0.73) 11.01					

Ō -mean values, SD – standard deviations, and variations: min – minimal value; max – maximal value. P – polar axis (µm); E – equatorial axis (µm); CL – colpus length (µm); CW – colpus width; EGT – exine thickness (µm); P/E – polar axis/equatorial axis ratio; Me – mesocolpium diameter; PLD – primary luminna diameter (µm); Mu – (primary) muri thickness; NoPL/25 µm^2 – number of primary lumina per surface area 5 µm x 5µm; SLD – secondary lumina diameter (µm); NoSL/PL – number of secondary lumina per primary lumina.

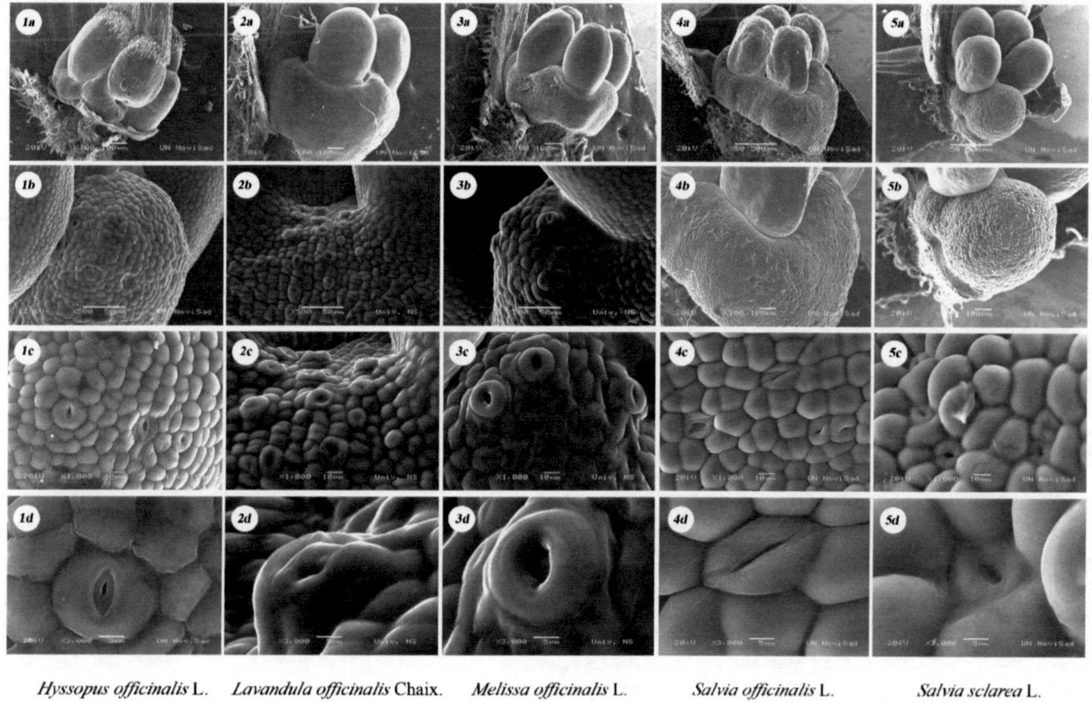

Figure 12. SEM microphotographs of the floral nectaries of five Lamiaceae studied: 1a-5a: Flower with sepals and petals removed, revealing the position of the nectaries. x 100 (1a,2a,3a); x 50 (4a,5a). 1b-5b: The nectary lobe facing the lower corolla lip. x 500 (1b,2b,3b); x 100 (4b,5b). 1c-5c: Detail of the nectary surface at the upper part of the biggest nectary lobe, showing irregularly distributed "modified stomata". x 1000. 1d-5d: Solitary modified stomata surrounded by 5 to 8 epidermal cells. x 3000.

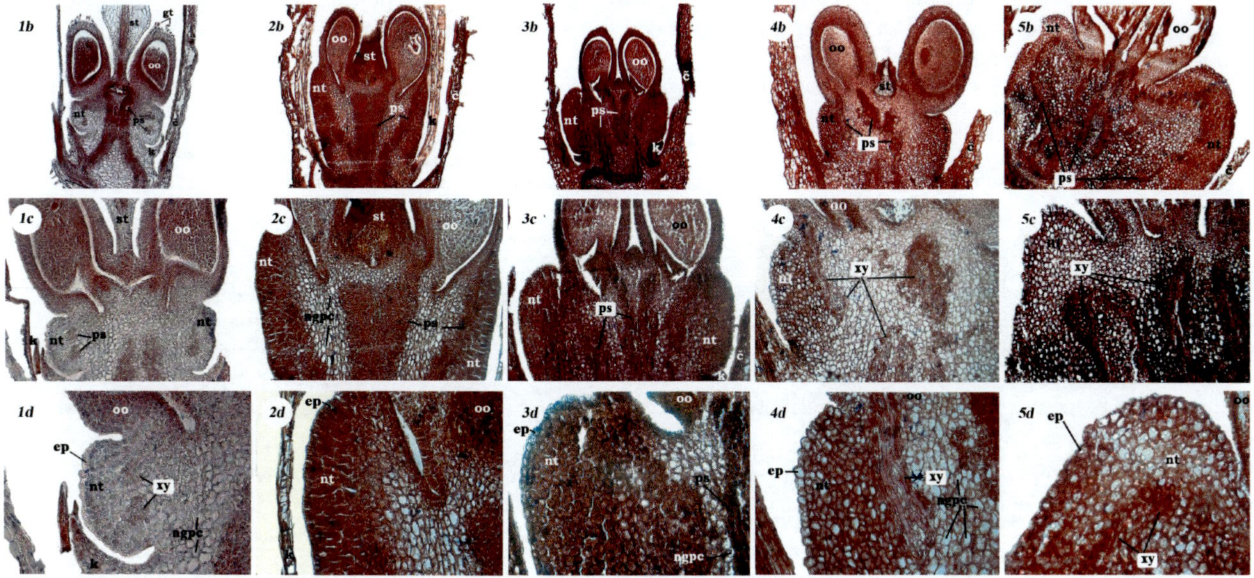

Hyssopus officinalis L. *Lavandula officinalis* Chaix. *Melissa officinalis* L. *Salvia officinalis* L. *Salvia sclarea* L.

Abbreviations: st- style; nt-nectariferous tissue; xy- xylem; oo- ovarian lobes; ps- vascular bundles; gt – glandular trichomes; k – petal; č– sepal; ngpc – non-glandular receptacle parenchyma cells; ep – nectary epidermis.

Figure 13. Anatomy of the floral nectaries of five Lamiaceae studied (Light microscopy). 1b-5b: Longitudinal section of the flower (receptacle and ovary) showing two opposite nectary lobes. x 100. 1c-5c: Secretory parenchyma cells clearly differed in size from the the non-glandular receptacle parenchyma cells, being smaller and densely packed. x 200. 1d-5d: The noticeable xylem touching or entering the nectariferous tissue. x 400.

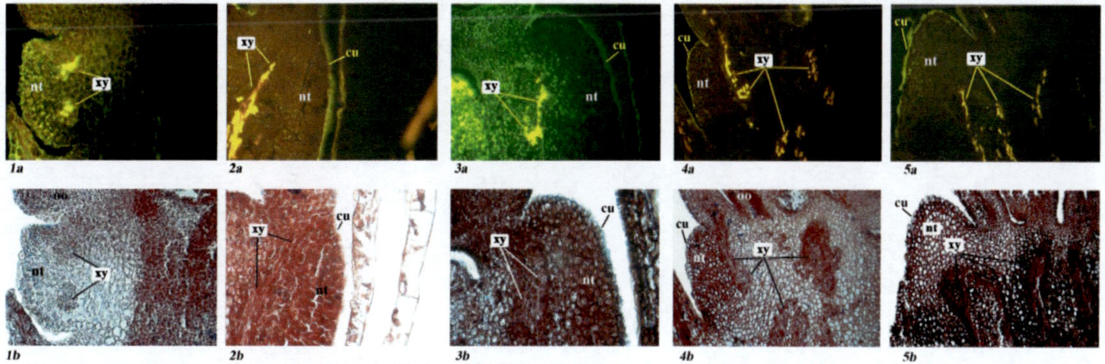

Hyssopus officinalis L. *Lavandula officinalis* Chaix. *Melissa officinalis* L. *Salvia officinalis* L. *Salvia sclarea* L.

Abbreviations: nt- nectariferous tissue; xy-xylem, cu – cuticle; oo – ovarian lobe.

Figure 14. Autofluorescence of nectary cuticle and lignified xylem. (*H.officinalis, L. officinalis, M. officinalis* - x 400; *S. officinalis* and *S. sclarea* - x 200).

Hyssopus officinalis L. *Lavandula officinalis* Chaix. *Melissa officinalis* L. *Salvia officinalis* L. *Salvia sclarea* L.

Figure 15. A flower base after removal of perianth (or perianth and androecium partly dissected), showing the nectary. The relatively large nectar droplets could be observed on the biggest nectary lobe of sage (4a) (Stereomicroscopy). Magnification: x 8 (1a); x 4 (2a); x 4 (3a); x 10 (4a); x2 (5a).

Morphoanatomical Characteristics of Floral Nectaries

Micromorphological Characteristics of Nectaries

Floral nectaries of all five examined Labiatae, with regard to their position in the flower, belong to the so-called toral nectary type, which have been described by Fahn (1979a; 1990). This means they are located on the receptacle, surrounding the ovary base in the shape of a protruding asymmetrical annular disc with four lobes. The nectary is slightly convex and its lobes alternate with four ovary lobes. SEM investigation demonstrated that the lobe which is closest to the lower corolla lip is noticeably more developed, i.e., bigger than the remaining three, in all species, except in *M. officinalis* (Figure 12 a, 15). In *S. sclarea*, the biggest lobe in the shape of a clearly separated protuberance, often elongated horizontally, parallel to the lower corolla lip (Figure 8 – 5a).

Measurements of the nectary diameter indicated that the largest gland has *S. sclarea* (2627 ± 95 μm), and the smallest *H. officinalis* (701 ± 46 μm).

As pointed out elsewhere for many other genera (Gulyás, 1967; Zer and Fahn, 1992; Mačukanović-Jocić et al. 2007; Bernardello, 2007 and references therein), the floral nectaries in Lamiaceae, generally, have a form of asymmetrical annular disc at the ovary base, from entire to four-lobed, sometimes with one lobe being more protruding than the others. Except for Lamiaceae, position on receptacle and general appearance of floral nectaries has been recorded in some species of other plant families, such as Boraginaceae, Convolvulaceae, Lythraceae, Ericaceae and Scrophulariaceae (Fahn, 1979a; 1990). The surface of the nectary consists of clearly separated, slightly protruding and irregular polygonal epidermal cells. SEM analysis revealed the presence of predominantly open stomata (Figure 12 – c, d), unevenly distributed mostly on the upper part (*L. officinalis*, *M. officinalis*) or the entire surface (*H. officinalis, S. officinalis, S. sclarea*) of the functional and, generally, the most developed nectary lobe. These typical structures, which are not able to contract their openings (Razem, 1998; Razem and Davis, 1999), designated as the so-called "modified stomata" have been recognized in a vast number of species (Rachmilevitz and Fahn, 1973, 1975; Eriksson, 1977; Zer and Fahn, 1992; Davis and Gunning, 1991, 1992, 1993). Also, a research of floral nectaries of rosemary revealed that modified stomata are present only at the top part of the biggest nectary lobe (Zer and Fahn, 1992).

The results from present study indicate that stomata are mostly individual or less often in pairs and irregularly distributed among epidermal cells lying in the same level (*H. officinalis* and *S. officinalis*) or slightly above (*L. officinalis* and *M. officinalis*). In *S. sclarea*, in which epidermal cells are more protruding, stomata are below their level (Figure 12 – 5c, 5d). Modified stomata are usually surrounded by 5 to 8 epidermal cells and are bigger than them in the hyssop, lavender and lemon balm, whereas in *S. officinalis* they are approximately equal in size (27 x 15 μm), and in *S. sclarea* they are much smaller (15 x 12 μm). They were not reported on any of the remaining three smaller lobes, except in *L. officinalis* in which, besides on the biggest lower lobe, two neighbouring lobes were found to be functional due to the presence of the stomata at their upper part.

Modified stomata, with slight variations in terms of size, position and density, are present also among other species that possess this type of nectar secretion. According to Bernardello (2007), nectar secretion through open stomata consisting of two guard cells that do not possess the property of nectar flow control, is the most common manner of nectar release. Even though most researchers agree that stomata which are connected with secretory tissue lost the property to be completely closed, there are other beliefs that in some cases they are able to regulate the size of the opening and nectar flow to a certain extent (Carr and Carr, 1987). There are opposing attitudes in the interpretation whether the amount of nectar in a flower is in correlation with the number of stomata. Certain researchers suggest that the volume of secreted nectar can be positively correlated with the number of stomata, whereas, there are those who claim that this is largely not the case, since not all present stomata contribute to secretion (Galetto, 1995; Gaffal et al., 1988; Petanidou et al., 2000). The first interpretation is in agreement with the present study, since in *S. sclarea* flowers, having the highest nectar potential, both the largest functional glandular surface and most stomata per surface unit were registered. Davis (2001) suggests that, while findings on the importance of number, density and distribution of stomata on nectary surface are gradually becoming richer, very little is still known about the molecular, genetic and environmental factors which control these characteristics.

Therefore, in the case of all five analyzed species, the nectar is transported from one to the next nectar-producing parenchyma cell towards intercellular regions, after which it is secreted outside through the stomata. The secreted nectar stays in the vicinity of the place of secretion, where it is accumulated in a reservoir formed by the corolla foundation, and relatively big nectar drops can be frequently noticed on lateral lobes (Figure 15).

Anatomy of Nectaries

The nectariferous tissue, observed and analyzed in the longitudinal sections by LM, consists of two hystologically differentiated regions, as also noted elsewhere (Fahn, 1988, 2000): a uniseriate epidermis overlying a glandular parenchyma (subepidermal secretory tissue) (Figure 13). Epidermis has been presented in a precisely defined layer of round or rectangular cells covered with a cuticle in all five species. The cuticle varies between species in terms of thickness, by keeping its continuity during the secretion phase. It demonstrates typical autofluorescence, when observed with an epifluorescence microscope (Figure 14). Epidermal cells are variable in shape and size among the studied Labiatae, possessing in most cases a convex outer wall. On average, in accordance with nectary size, the largest epidermal cells have been registered in *S. sclarea* (21.48 x 20.30 μm), and the smallest in *H. officinalis* (10.35 x 10.45 μm).

Beneath the epidermis there is the glandular parenchyma, also called specialized parenchymatous tissue (Fahn, 2000) or nectary parenchyma as suggested by Nepi (2007), consisting of between six and ten layers of polygonal, relatively isodiametric cells. Secretary cells are smaller than the neighbouring non-glandular parenchyma cells of the receptacle, densely packed and richer in cytoplasm, as pointed out earlier (Nepi, 2007), forming glandular tissue easily noticeable using LM (Figure 13-c,d). In terms of size and shape of glandular tissue cells, differences were recorded among the species studied. The secretory cells of *L. officinalis*, deviate the most from the polygonal shape. They are extremely densely packed, narrowed and elongated (on average 11.81 x 5.98 μm) (Figures 13-2d). The largest secretory cells, directly proportional to the size of nectaries are recorded in *S. sclarea* (19.77 x 18.89 μm) (Figure 13 – 5d).

According to Pacini et al. (2003), characteristics of nectariferous tissue can vary depending on nectary type and manner of secretion. Also, the presence of several layers of relatively small and densely packed secretory cells with densely staining cytoplasm has been previously described in most floral nectaries in the secretory stadium (Fahn, 1979b; Link, 1992; Nepi et al., 1996; Davis et al., 1988a; Vesprini et al., 1999; Razem and Davis, 1999).

Beneath the nectary parenchyma, there is the subnectary parenchyma as proposed by Nepi (2007), which has also been designed as subglandular parenchyma (Durkee, 1983). Nectaries appear to be supplied by their own vascular elements constituting a branch of the main vascular bundle of receptacle, through a band of large, tightly packed subglandular parenchyma cells. According to Nepi (2007), vascular bundles may be found in the nectary or subnectary parenchyma. Although nectaries are reported to be vascularized primarily by phloem only or by xylem and phloem equally (Nepi, 2007), xylem elements alone has been visible in the longitudinal sections of flowers, reaching the nectary parenchyma. Xylem has been noticed, close to the nectariferous tissue (*L. officinalis* and *S. officinalis*), or inside it (*H. officinalis, M. officinalis* and *S. sclarea*). This has been confirmed in micrographs made on a fluorescence microscope, thanks to the autofluorescence property of xylem elements (Figure 14). It cannot be said with great certainty for any of the analyzed species which type of nectary vasculature is present (phloem and xylem or only phloem) since phloem elements were not noticeable in the area of glandular parenchyma cells.

Sometimes the sieve tubes do not enter the secretory tissue but only border it (Fahn, 1988; 2000). Even though these main vascular bundles are collateral and consist of phloem and xylem, which has been demonstrated by earlier research (Fahn and Shimony, 2001), in some species of the Lamiaceae family, e.g., *Rosmarinus officinalis, Salvia judaica* and *Satureja thymbra* only phloem has been registered in secretory tissue, and in others, e.g., *Phlomis viscosa, Salvia fruticosa* and *Stachys aegyptiaca* – both xylem and phloem (Dafni et al., 1988). Zer and Fahn (1992) have recorded when studying floral nectaries of rosemary that only phloem supplies the inner part of each lateral nectary lobe.

Seeing that nectar, in terms of its chemical composition, represents a modified phloem sap whose constituents are products of photosynthesis (Fahn, 2000), it can only be assumed that among species in which xylem is perceived inside nectary tissue, probably elements of phloem exist, and in species in which xylem was perceived only nearby, any of the two types of vasculature can be perceived. Eventually, the reason for the inability to notice phloem in the longitudinal sections of analyzed flowers can be the limitation of the applied method.

Literature data show that the process of photosynthesis in the function of nectar production can take place in the gland or other plant parts (Nepi et al., 1996; Vesprini et al., 1999). In the first case, sugar is produced in the nectary itself, in which it is stored in the form of starch for only a few hours, after which it undergoes hydrolysis during the night for producing nectar. In the other case, the nectary parenchyma serves exclusively to store starch (in amyloplasts), which is hydrolyzed immediately before the secretion begins (Razem and Davis, 1999). Still, according to Nepi (2007), the main photosynthetic activity takes place in subnectary parenchyma where a greater quantity of chlorophyll is located.

Comparative Analysis of Nectar Production

Comparative analysis of nectar production in five Lamiaceae species has shown that *S. sclarea* has the highest total daily nectar amount per flower (on average 1.003 ± 0.344 μl), the highest summary diurnal nectar amount per flower (on average 2.288 ± 0.766 μl), and the highest average nectar secretion rate (0.152 ± 0.049 μl/h). The lowest 24-hour nectar accumulation has been recorded in *M. officinalis* (0.107 ± 0.055 μl), the lowest amount secreted during daytime (average 0.236 ± 0.178 μl) and the lowest nectar secretion rate (0.015 ± 0.012 μl/h). Therefore, lemon balm can be considered the slowest nectar producer in the given environmental conditions (Figures 16, 17 and 18).

Results clearly show that in all species, the summary diurnal amount, even though it relates to a time period which is twice as short (during 12 hours), is twice as high in comparison with the 24-hour sample indicating that the intensity of this physiological process is different in visited and unvisited flowers. Even though earlier research confirmed that secretion is resumed if pollinators periodically collect nectar from flowers (Boëtius, 1948; Wykes, 1950; Raw, 1963), certain authors believe that new initiation of nectar secretion after its removal in experimental conditions can lead to a wrong assessment of the total amount of nectar a flower can produce during the day (Cruden et al., 1983).

However, the nectar potential of a flower on a daily level shall be assessed more accurately if it relates to the summary amount of nectar sampled several times during the daytime, which simulates the periodic pollinator visits, and not to the amount of nectar sampled once after 24 hour daily accumulation. The answer is physiological and lies in the fact that the flower usually starts to secrete nectar somewhat before the foraging activities of its co-evolutionary pollinators with a constant secretion rate (Witt et al., 1999; Nepi et al., 2001) until a critical amount is accumulated, and then the secretion stops (Hainsworth and Wolf, 1972a; Stiles, 1975; Wolf, 1975).

Discontinued nectar production during the day, when the secretion phase is followed by a short or long break, as well as a complete cessation of this process can arise due to some of the following reasons: reaching of the maximum, during the inactivity phase of the pollinator (e.g., overnight for daily active pollinators, and vice versa), if pollination has been successfully performed, or because of unfavourable microclimatic conditions (Cruden et al., 1983 and references therein).

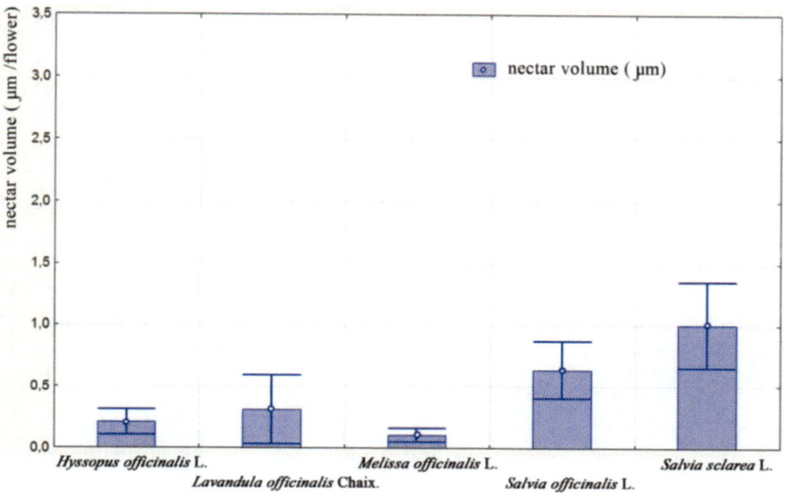

Figure 16. Average total daily nectar production per flower in five Lamiaceae species during the anthesis period in 2004.

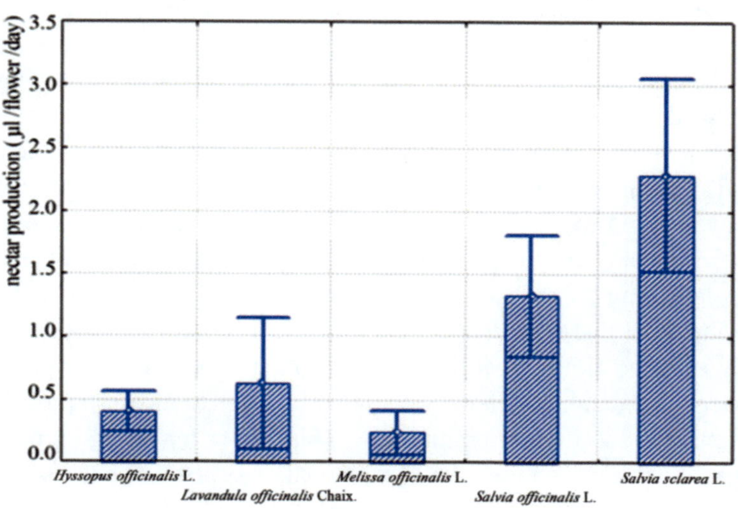

Figure 17. Diurnal summary nectar amount per flower in five Lamiaceae at the average level during flowering period.

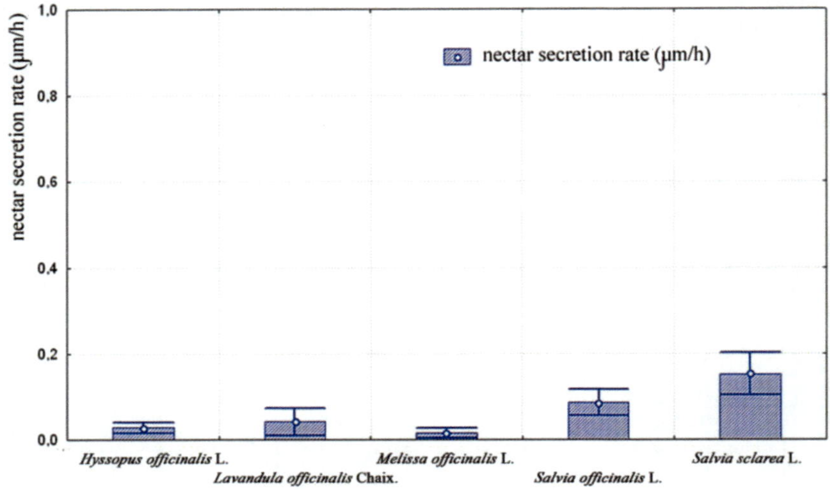

Figure 18. Average nectar secretion rate in five Lamiaceae species during the anthesis period 2004.

Therefore, the 24-hour accumulation in the present research is lower than the summary diurnal amount, because quickly after attaining critical amount, the flower stopped its secretion. The fact that it kept producing on a necessary rational minimum, and offering an attractive and tasty meal in portions is justified from the point of view of stopping unnecessary energy wasting. On the other hand, some research have shown that if one pollinator visitation is enough for a flower to be efficiently pollinated, removing the accumulated amount will not incite it to produce and secrete again (Cruden et al., 1983).

However, in our research, the periodical removal of nectar from the flower using microcapilary led to the continuation of secretion, albeit at a lower intensity, according to the supply and demand principle. Pacini and Nepi (2007) pointed out that re-activation and increase of total nectar production in repeatedly sampled flowers can be related to their need for a few pollinator visits to deposit enough pollen on the stigma in order to increase probability, i.e., to ensure successful pollination. This rule regarding production resumption after nectar removal has been noted, except for floral, also in the functioning of some extrafloral nectaries, but with different ecological explanation (Heil et al., 2000).

By analysing the nectar secretion rate (nectar amount secreted in an hour) and connecting it to the behaviour and activity of flower visitors, Cruden et al. (1983) proposed a general division of flowers in three groups: slow, fast and "super" producers, stressing the differences from the morphophysiological standpoint. Namely, the beginning and intensity of nectar secretion are in correlation with the feeding activities of co-evolutionary pollinators, and also reflect, in some cases, the level to which food source has been protected from the so-called nectar thieves (visitors that irregularly approach flowers by finding a shortcut to the reservoir of sweet liquid, without performing pollination). The plants which have been designated as slow producers (secrete 5-10% of their maximum accumulation per hour) have a slow initiation of secretion process and very well protected accumulated nectar (reinforced corolla or calyx). Contrary to these, there are flowers that belong to fast nectar producers (secrete 22-68% of their max accum. per hour), among whom the secretion is initiated later, and the nectar is protected more weakly. “Super producers” (secrete two or three times as much as fast producers) are species whose flowers are also visited by the thieves, half an hour or one and a half hours before the activity of their pollinator starts.

The flowers of the final group, thanks to the high rate of nectar secretion, regardless of the stolen amount, offer their pollinators enough nectar at the time of their activity. The high intensity of nectar secretion is interpreted as adaptation that enables these plants to be competitive for pollinators in spite of the regular loss of nectar due to theft. Having in mind the classification proposed by Cruden et al. (1983), lemon balm could be classified in the first group of slow producers (Figure 18).

Five Lamiaceae species secrete nectar continuously during all stadiums of the flower ontogeny. The beginning of secretion in most cases has not been recorded immediately after the bud opening, but somewhat later, during which time it shrinks and stops when the first signs of corolla wilting are noticed. The flowers of five species are different from each other according to the phase of ontogenetic development in which they produce the maximum amount of nectar per flower: hyssop – on the first day (four days lifespan), lavender – day two (seven days lifespan), sage – on day three (four days lifespan), clary sage – day one and day three (four days lifespan) (Figure 19).

It has been noticed that the lifespan of marked flowers from which nectar was collected, was shorter compared to the covered non-sampled flowers in which the approach of pollinators was not allowed.

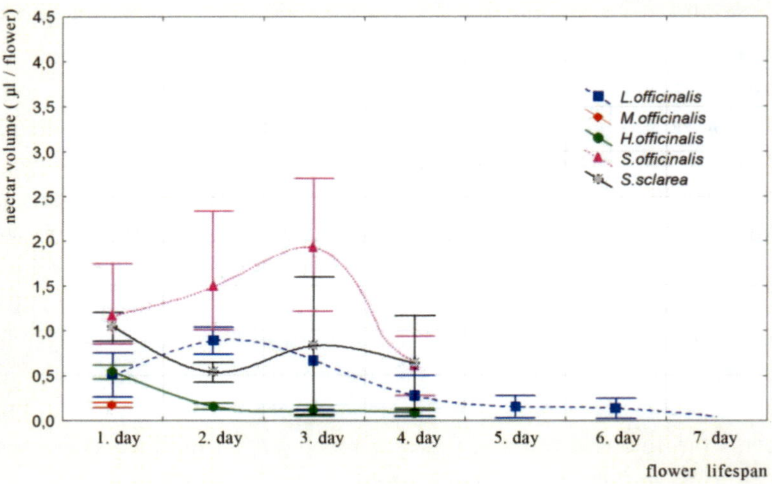

Figure 19. Nectar production during flower ontogeny in five Lamiaceae species.

Considering that the flower lifespan of a lemon balm is only around twenty hours, the nectar secretion during the ontogenetic development corresponds to the diurnal dynamics of nectar secretion. Namely, there is no nectar in the lemon balm flower immediately after the bud is opened. The secretion starts a bit later, reaching its maximum in 3 to 5 hours. The process of floral nectar secretion coinciding to the pollination phase, and usually starts immediately after the bud opening, following a certain rhythm which mostly depends on external factors including pollinators, and then its intensity drops as the flower gets older.

After the pollination is performed, in most cases the shrinkage takes place or the nectar is no longer secreted. As pointed out elsewhere (Vogel, 1983), in rare occasions (*Lamium* sp.), the secretion is continued even after the corolla falls off. Mostly, just open and old flowers contain less nectar in comparison with those in full maturity (Vogel, 1983). In some cases, most nectar is produced in the bud phase (Torres and Galetto, 1998). According to Fahn (1949), in flowers of *Calllistemon lanceolatus* and some species of the genus *Citrus*, whose life period is 3-5 days, maximum secretion usually occurs on the first day, which is also the case with hyssop and clary sage in the present study.

Based on the comparative analysis of the diurnal dynamics of nectar secretion in the given microclimatic environmental conditions, it has been established that all analyzed Labiatae flowers secrete nectar continuously during the day (Figure 20).

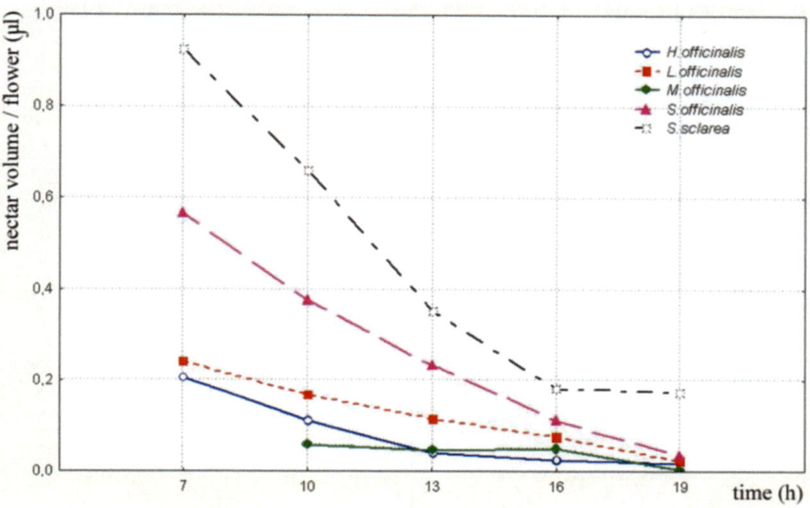

Figure 20. Diurnal dynamics of nectar secretion in five Lamiaceae species (average values during nectar collecting days in 2004).

The high values of standard error in certain measurements can be explained with the minor amount or a complete absence of nectar in certain flowers in certain time intervals. This particularly relates to the flowers of lavender and both sages. Two patterns of the diurnal dynamics of nectar secretion with a difference in time necessary to achieve the secretion maximum have been defined. In the first model, which is typical for *H. officinalis*, *L. officinalis*, *S. sclarea* and *S. officinalis*, secretion was most intense in the mornings (between seven and ten a.m.) with the decreasing tendency during the day. In the second model, which was recorded in *M. officinalis*, in which the production during the day is maintained on the lowest level, a weak secretion peak was reached in the afternoon (16 ^{h}). Species whose flowers behave according to the first model have a minimum amount of nectar at a time when the lemon balm has a slightly higher secretion. The reason for difference in the secretion dynamics could lie in the fact that the flowers of lemon balm mostly open around 10^{h} (approximately 95%) and seeing that they are poorly visited and not competitive with other species, they start to intensify secretion later. The reason for the great amount of nectar secreted in the morning in the flowers of other species (which does not coincide with the period of most intensive bee activity, except in the case of hyssop; Figure 22) is the early accumulation of nectar which preceded the first sampling, caused by the covering of plants during the previous day.

The asynchronous rhythm of nectar production is frequent in habitats in which competition for pollinators exists.

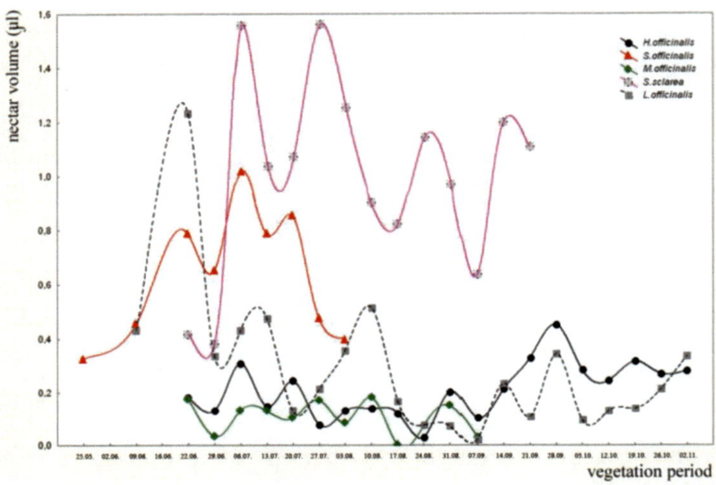

Figure 21. Seasonal dynamics of nectar secretion in five Lamiaceae species during the anthesis period in 2004.

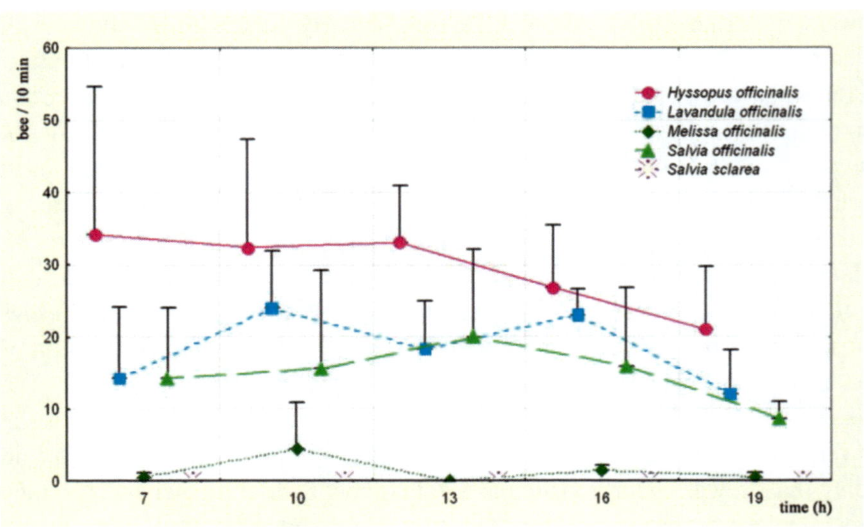

Figure 22. Diurnal dynamics of honey bee visit frequency during the nectar collecting days (mean values are shown) related to the entire observation period in 2004.

This phenomenon, which is of adaptive significance due to the tendency of reducing the competitive relationships, was recorded by Kleber (1935) in three Asteraceae species blooming simultaneously, whose flowers, pollinated by bees, intensified nectar secretion at different times. Considering the behaviour of pollinators, *Erigeron simplex*, *Senecio crocatus* and *Erigeron peregrinus* reach the maximum amount of nectar at different times of the day (mid-morning, late morning and early afternoon).

However, unlike mentioned case, in most studied Lamiaceae plants, the timing of this process is fairly synchronous, since the maximum secretion occurs early in the morning (except for lemon balm), and some other morphophysiological characteristics would have to be of crucial importance in the attraction and preferential foraging behaviour of bees.

Nectar production and secretion in the function of time has been studied a lot in several plant families. For example, changes in nectar production and composition during the day among species from the families Boraginaceae and Brassicaceae were described by Corbet (1978). By studying the diurnal and the seasonal dynamics of nectar secretion in the Lamiceae and Caryophyllaceae families, some authors concluded that this process differs between species as well as within the same species (Mačukanović and Blaženčić, 1998; Mačukanović-Jocić et al., 2004; Witt et al., 1999).

Dynamics of nectar secretion has been probably studied in most detail in the Rosaceae family and in the subfamilies Prunoideae (Orosz- Kovács, 1992; Orosz- Kovács et al., 1989, 1992, 1995) and Maloideae (Scheid-Nagy Tóth, 1991, 2000; Orosz- Kovács et al., 1994; Szabó-Mühlenkampf, 1994; Farkas and Orosz- Kovács, 2003).

By comparing five Lamiaceae species with three forest species belonging to the same family, namely *Lamium maculatum*, *Lamiastrum galeobdolon* and *Ajuga reptans*, which grew in natural and experimental habitats (Mačukanović-Jocić et al., 2004), completely opposing models of diurnal secretion activity were recorded. Even though in forest species the intensity of nectar secretion tends to grow during the day, the results showed the expected differences in the intensity and dynamics of nectar production and in the sugar concentration in different habitats, which can mostly be attributed to the synergistic influence of air temperature and humidity.

These microclimatic parameters express a dominant, direct or indirect effect on the rate of secretion and nectar quality. When the air temperature exceeds the optimum level, the secretion decreases, and the extremely low and high temperatures stop this process (Lazarov et al., 1976). According to Vogel (1983), this process (qualitatively and quantitatively) is, above all, under the influence of temperature. Temperature conditions which define the commencement, maximum, cessation and the optimum intensity of this process differ between different species in accordance with the climate they grow in.

Air humidity most of all affects the viscosity, i.e., sugar concentration of secreted nectar, which is more expressed in the event of exposed and unprotected nectaries (Pacini and Nepi, 2007).

Seasonal dynamics of nectar secretion in five Labiatae is connected to the effect of climate on seasonal occurrence of these plants, and on the beginning, course and duration of the blossoming phenophase. It has been recorded that secretion is continuous during the flowering period (Figure 21). Discontinuity in nectar production occurs in these plants grown in plantations if they have two peaks of more massive blooming, which is a consequence of mowing or planting at different times (in the case of *S. officinalis, H. officinalis, S. sclarea*). Also, the termination of secretion was caused by a consecutive series of cold or overly warm days with temperatures above average. Secretion was most intensive in the period of massive blooming of each species, and subsequently, with a smaller or greater fluctuation, it decreased more mildly or more intensively (*L. officinalis, S. officinalils* and *S. sclarea*) or maintained at the same level (*M. officinalis*) towards the end of the vegetation period.

In hyssop, a small increase of the nectar production intensity in early autumn has been recorded. In days when air temperature were not in the optimum range (e.g., July 20 – average and maximum daily temperature above 30°C, and on 12 October – average and maximum daily temperature below 10°C), the amount of nectar produced in flowers was lower, which proves that microclimatic conditions affect the intensity of this process.

In a similar research, conducted by Nepi et al., (2003) in *Linaria vulgaris*, which is characterized by a long period of inflorescence (VI-XI), the amount of nectar per flower and sugar concentration varied a lot during the season, also, depending on environmental parameters.

Seasonal changes in relative air humidity and temperature are in accordance with the moderately continental climate of Belgrade and its surroundings, which has been confirmed by microclimatic measurements performed in the field, and by results provided by the Republic Hydrometeorological Service of Serbia. The effects of microclimatic parameters, especially air temperature and humidity, on the nectar production and secretion have been confirmed in the present research, especially when the values exceeded the optimal range for this physiological process. Namely, with the increase of air temperature and the drop of relative humidity from the morning to the early afternoon, the intensity of nectar secretion also drops in most species. The temperature range in which the nectar production process took place was between 10 and 30°C with the optimum between 16 and 25°C, which, according to Škenderov and Ivanov (1986) and Lazarov et al. (1976) is typical for most species which grow in a moderate climate.

The present study confirmed that nectar production is a complex physiological process controlled by genes, even though it is conditioned by numerous endogenous and exogenous factors (Walker et al., 1974; Cruden and Herman, 1983; Mitchell, 2004; Leiss and Klinkhamer, 2005). The complexity of this process, specific for the flowering plants with floral nectaries, is reflected in the existence of their own secretion rhythm during the day, the flower ontogeny and the flowering phenophase (having in mind that the commencement, termination and rate of secretion as well as the minimum, maximum and total daily nectar volume per flower, are different among species) which, however, is highly subject to changes under the influence of abiotic and biotic environmental factors (Nicolson and Nepi, 2005; Petanidou, 2007).

The existence of a certain secretion pattern, i.e., quantitative and qualitative properties of the nectar of each plant species can be conditioned by its specific internal factors which refer to the characteristics of flowers, i.e., inflorescence (shape, size, position, development phase, number and distance between flowers) and nectaries (position and size of glands, presence of modified stomata, etc.) (Gottsberger et al., 1990; Petanidou et al., 1996; Davis et al., 1996; Ricciardelli D'Albore and Intopa, 2000). Pacini and Nepi (2007) reported the existence of intraspecific variability of nectar features within individual flowers; among flowers of the same plant, between plants of a single population and between different populations. Also, soil properties, air composition, sunlight intensity, precipitation and other abiotic factors affect this process (Cruden et al., 1983; Škenderov and Ivanov, 1986; Freeman and Head, 1990; Wyatt et al., 1992; Jakobsen and Kritjansson, 1994; Lake and Hughes, 1999; Rusterholz and Erhardt, 1998).

Pollinators with different energy demands and the nectar thieves have been classified as a separate category of external factors. The amount of nectar in a flower in the function of time, as the consequence of secretion, and in some cases reabsorption and the qualitative nectar characteristics, reflect the adaptive response of plants to the foraging behaviour of pollinators they co-evolved with most of all (and also with undesirable flower visitors) in combination with abiotic environmental factors (Corbet, 2003; Pacini and Nepi, 2007; Nicolson, 2007).

STUDY OF FLOWER VISITORS

All of the species studied are insect pollinated. With respect to pollinating insects, three types of pollination were represented: melitophily, myophily and psychophily. Bee visitors included honey bee (*Apis mellifera* L.) and bumblebees (*Bombus terrestris* L./*lucorum* L., *Bombus hortorum* L., *Bombus pascuorum* Scopoli and *B. lapidarius* L.) that collected nectar and pollen intensively from hyssop, lavender and sage, and carpenter bee (*Xylocopa violacea* L.) which was the only bee species able to collect nectar legitimately from *S. sclarea*. Of a total of insect visits recorded, most were made by honey bees and than by bumblebees. Other insects, making an average less than two visits per plant in 10 minutes, were flies, butterflies and solitary bees.

Fly visitors included pollen and nectar-collecting hoverflies and bee flies. A couple of the most frequent hoverflies (Syrphidae), identified as *Eristalis tenax* L. and *Episyrphus balteatus* De Geer, were both attracted by *L. officinalis*. *Bombylius major* L. (*Bombyliidae*) was seen exploiting lavender and hyssop corolla tubes. Concerning these two plant species, several cabbage butterflies (*Pieris brassicae* L., Pieridae) were usually found together on the same plant and the individuals remained a long time in the area feeding slowly. The adults of *Macroglossum stellatarum* L. (Sphingidae), have been observed as very active during daytime, but most of them have usually been seen from 12.00 till 19.00 h. The major visitors identified to species and visitor counts are listed in Table 3.

Honey Bee Visits

The results of the observations performed during the research period demonstrate varying levels of attraction of five Lamiaceae species for the honey bee.

Comparative analysis of the honey bee visit frequency to the species analyzed, at the average level for the entire research period, shows that the most frequently visited were flowers of *H. officinalis* (30.28 bees/10 min) and the least those of *M. officinalis* (1.4 bees/10min) (Table 3). Considering the entire research period (Figure 22), the following observations were made: continuous and massive flower visitation by honey bees during the daytime was observed in the case of *S. officinalis*, *H. officinalis* and *L. officinalis,* discontinuous and sporadic in the case of *M. officinalis* displaying an obvious lack of interest for plant, and finally, flowers of *S. sclarea* were never visited.

Table 3. Flower visitors and their number in 10 min per plant during nectar collecting days

Plant species	Flower visitor	Frequency of visitors (No/10 min/plant)
H. officinalis	*Bombus terrestris/lucorum*	6.17
	B. lapidarius	5.16
	Apis mellifera	30.28
	Pieris brassicae	sporadically
L. officinalis	*Bombus terrestris/lucorum*	4.23
	Bombus pascuorum	3.10
	Apis mellifera	18.33
	Pieris brassicae	sporadically
	Macroglossum stellatarum	1.30
	Episyrphus balteatus	1.90
	Eristalis tenax	0.90
	Bombylius major	1.97
M. officinalis	*Apis mellifera*	1.40
S. officinalis	*Bombus terrestris/lucorum*	sporadically
	Apis mellifera	14.93
	Bombus pascuorum	14.50
	Bombus hortorum	3.15
S. sclarea	*Xylocopa violacea*	0.13

Data represent average values for the entire investigation period.

The Percent of Visited Flowers

The percent of visited flowers with respect to the total number of open flowers on a plant was calculated on the basis of the total number of bees and the average number of flowers visited by a single bee in ten minutes. The comparative analysis of the percentage of flowers visited in the ten minutes at the average level during the research period shows that the most visited flowers were those of *L. officinalis* (61.12%), followed by *H. officinalis* (43.67 %) and *S officinalis* (18.22 %), while the least visited flowers belonged to *M. officinalis* (2.02%). Flowers of *S. sclarea* were completely unvisited.

Without taking into account the other parameters that affect the attractiveness, and having regard to the observation that hyssop was the most frequently visited, a higher percentage of visited lavender flowers can be interpreted by a generally smaller number of flowers per inflorescence and per plant in some phases of the blooming period. This was also noted in the period when the lavender population was mostly in the fruiting phenophase, even though it blossomed occasionally, and for a certain time after mowing (in order to use drugs for medicinal purposes), when a small number of shoots flowered afterwards. The second reason is the longer retaining of bees on lavender in comparison with hyssop, whose flowers the bees were quickly leaving. Long-term retention of the few bees in a lavender plant could be explained by the strategy of bee's behaviour, which, regardless of the better offer of other plants in its surrounding, aims to collect the necessary amount of nectar to satisfy its own needs and the need of the colony while spending as less energy as possible. Flying over small distances between closely distributed flowers of the same species saves time and energy. Also, by unmistakeable repetition of the procedure to approach the familiar flower and reaching the nectar reservoir, energy is not needlessly lost on the process of learning how to feed on morphologically different flowers.

Therefore, the smaller number of open flowers and longer retention of bees on a plant (more flowers visited by a single bee) caused a higher percentage of visited lavender flowers in the observed time interval, i.e., it increased the odds that a single flower is visited more than once.

In general, the percentage of visited flowers in hyssops, lemon balm and sage is relatively low compared to similar research performed in other melliferous species of the family Lamiaceae (Dafni et al., 1988) in which this percentage amounts to around 70% in the case of *Corydothymus capitatus* e.g., or above 100% in *Rosmarinus officinalis*. Hence, while in *Rosmarinus officinalis* each open flower was exploited by bees more than once in ten minutes, in the present study the most visited lavender flower was exploited only 0.6 times on average in the 10 minutes. However, as well as in the aforementioned research, a small percentage of visited flowers in *Melissa officinalis* was recorded in the present study.

Most of all, this can be explained by large plantations in vast production area in which these and many other melliferous plants are grown (*Facelia, Ocimum, Calendula, Satureja, Origanum*...). This implies a significant offer in terms of bee pasture, compared to relatively small number of bees. It is also well known that bees land on flowers of a certain species as long as it is at the peak blooming phase, after which they searching for another source of floral nectar.

Five Lamiaceae species were visited by bees during their flowering period in the production field of the Institute for the Medicinal Plant Research, from mid-May to September. However, in this locality, the microclimatic conditions (short sunny intervals, lower daily maximum temperatures, and fertile soil) are different from habitat conditions in which these plants grow naturally.

The conditions of plantation growth in the imposed environment affect the shortening of the flowering period, which in favourable conditions can last until October (especially in the case of hyssop and lavender). Also, planned mowing of these species, after a massive flowering for medicinal purposes, disturbs the flowering continuity, which affects bee visitations.

Photo: M. Mačukanović-Jocić and G. Jocić.

Figure 23. *Macroglossum stellatarum* pollinating nectar poor flowers of *Lavandula officinalis.*

ATTRACTIVENESS OF LAMIACEAE SPECIES COMPETING FOR POLLINATORS

The investigation of attractiveness and melliferous potential of five Lamiaceae species revealed that the most of the observed pollinators were comparatively rare, indicating the flowers to be attractive almost exclusively for hymenopterans of the Apidae family. It has been noticed that *H. officinalis*, *L. officinalis* and *S. officinalis* are "generalists" in terms of multiple choice insect pollinators. The main visitors, predominantly to the blue flowered species rich in nectar, were honey bees and species of bumblebees.

The less frequent flower visitors included solitary bees, small and large hoverflies, bee flies and butterflies. Syrphidae and Bombyliidae are usually generalists in their flower preferences and have been mentioned as pollinators or regular visitors of thousands of species of flowering plants, including Lamiaceae (Herrera, 1988; Delgado, 2000).

Butterflies made relatively few daytime visits discontinuously during the investigation period. The most frequent visitor was the hawkmoth

Macroglossum stellatarum, particularly found on *L. officinalis* (Figure 23) whose flowers were not the richest in nectar among the investigated species, despite the conclusion of Kelber (2003) that adults, having high energy demands, are fond of flowers with lots of nectar.

Given that, according to Kelber (2002), this moth is able to associate flower colour, size and pattern with the presence and absence of food, it is obvious that other attractants, rather than the amount of nectar, determined its flower preferences. So, *L. officinalis* was attractive for high-energy insect pollinators, but was also pollinated by small bee visitors. Results based on this observation concerning correlation between nectar production and insect visits, mostly disagrees with those of Cruden and Toledo (1977) and Cruden et al. (1983), who suggested that quantitative and qualitative characteristics of floral nectar correspond to the energy demands of consumers, indicating that flowers pollinated by sizable visitors should have higher nectar potential over those pollinated by small ones.

Nevertheless, these plants are not dependent on flies and butterflies for pollination. The species of *Bombus,* depending on proboscis length as a morphological indicator of resource utilization, were noticed getting nectar from blue flowered species. Interdependence between proboscis length and the time bees spend on flowers with different corolla tube length has already been widely discussed (Herrera, 1989; Inouye, 2004).

Fairly common bees, Bombus terrestris and B. lucorum, whose workers were almost indistinguishable, have been seen together, pollinating the flowers with smaller corolla tube length such as hyssop and lavender. *B. lapidarius,* having comparatively short tongue, preferred inflorescence of hyssop made up of many small flowers, while *B. pascuorum* with a medium length tongue, was often found foraging on sage and lavender. Such observations support Comba et al. (1999) argument that bee visitation of moderately long flowers also in some other Lamiaceae species such as *Salvia pratensis, Lamium album* and *Stachys slyvatica* (mean corolla depth from 8.5 mm to 11.37 mm) indicates the morphological compatibility between these plants and their pollinators.

In lieu with the predictions that the five investigated species are predominantly bee-pollinated, flower morphology has been discussed in relation to honey bee (*Apis mellifera* L.) attraction as the most frequent and the most effective pollinator observed. Despite the fact that the honey bee is polylectic, and cannot specialize on particular species, but must learn how to handle the actual flowers (Schmidt and Johnson, 1984), this investigation showed that it has never visited *S. sclarea,* whose exclusive pollinator was the solitary carpenter bee *Xylocopa violacea*. The sizable flowers of *S. sclarea* with high positioned fertile parts beneath the arcuated upper corolla lip and baggy incavation of lower lip, represent a morphological constraint for honey bee pollination.

The foraging choices of honey bees are affected by many factors, such as floral scent, colour, etc. It was observed that the most adapted flowers for honey bee pollination, considering the lower lip size as a landing platform and the largest opening of the corolla tube, are those of *S. officinalis.* From the visitor's point of view, the flower of *M. officinalis* is the least accessible, due to the corolla tube length, small diameter of its opening (average 2 mm), sigmoid corolla, non-conformity of lower lip towards landing, and easy separation of calyx and corolla upon minimal touching. By its morphology, hyssop flower is highly adapted to honey bee pollination, with the shortest corolla tube (average 6.76 mm) and nectar storage at a depth suitable for the proboscis length. This species secrete less nectar per flower than lavender, but its profitability is increased because the flowers are grouped in a long dense inflorescence, reducing interflower travelling costs for foraging insects.

The correlation between corolla tube length and flower attractiveness for honey bees was not established, but this trait could be significant. There is an opinion that the honey bee never visits flowers with corolla tubes longer than 10 mm, because its proboscis is around 6.5 mm (6.2 - 6.7 mm) long, disabling it to rich the nectar (Baker and Baker, 1983; Corbet et al., 1995).

Comparing the length of proboscis to that of the corolla tube leads to the conclusion that *H. officinalis* and *L. officinalis* should be the most attractive. However, the difference in the level of their attractiveness was recorded, meaning that in the same time interval, approximately 30 honey bees visited one plant of hyssop, and 18 one plant of lavender.

On the other hand, *S. officinalis* and *M. officinalis* have similar corolla tube length (11 mm and 10.5 mm) but the flowers of the former were highly, and of the latter poorly attractive.

Flower size had no crucial influence to the bee attraction if discussed separately, but naturally it does in combination with other characteristics such as colour, shape and the interfloral distance and inflorescence size. It could be supposed that this trait contributes to the size of coloured object registering in the honey bee`s field of view.

Besides the floral structure, flower colour and odour, nectar guides and nectar characteristics are also voted as attractiveness factors (Harborne, 1988). According to the data from the literature and on the basis of our subjective assessment of scent intensity, all the investigated species are considered to be very aromatic. Plant scent, deriving from volatile essential oils, is reputed to be the most important factor of biochemical communication (attractant or repellent) between plants and animals (Harborne, 1988). Insects usually have a highly susceptible sense of smell and are able to detect the essential oil at a substantial distance from the plant. Attracted by the scent, they are approaching the flower, whereby color and shape help them in the orientation (Jančić et al., 1995). Nevertheless, some Lamiaceae flowers, like those of *Melissa,* are scentless, therefore, odour from the whole plant attracts over longer distance, and thereafter the inconspicuous scent of the corolla precisely leads to the nectar (Beker et al., 1989; Jančić et al., 1995). Based on some earlier investigations, corolla of *M. officinalis* contains small amount of essential oil (0.002%) of complex composition with dominance of β-cariofilen (Schultze et al., 1989) and the dried lemon balm leaves contain 0.32% essential oil predominantly of aromatic and polyphenolic constituents (Carnat et al., 1998). Umeljić (1999) characterized *M. officinalis* as a very good melliferous plant, which is strongly contradicted by our results that revealed its low nectar production, upholding some earlier investigations which pointed out that the plant is poor food source for the honey bees (Dafni et al., 1988).

At shorter distances, a pollinator additionally receives a visual signal. Not all the colours are equally attractive to honey bees. Bee flowers have brightly coloured petals that appear to the human eye as blue or yellow colours (Fritsch, 1950).

Apis mellifera very rarely or never visited mainly white coloured flowers of *M. officinalis* and *S. sclarea*, although Harborne (1988) mentioned that the bees are sensitive to UV-absorbing flavones and flavonols present in white corollas, being instrumental in bringing about the fascinating pollinating guide patterns. Nevertheless, good evidence of significance of corolla colour as a factor of attractiveness was that the most frequently visited were the blue flowered species (hyssop, lavender and sage).

As the bee alights on the flower, it is guided towards the nectar reservoir by visible or invisible honey guides on the petals (Harborne, 1988). The role of floral guides in the process of pollinator orientation have been investigated experimentally (Waser and Price, 1985). Distinctive colour patterns, as visible honey guides, have been evidenced in the present study only on *S. officinalis* flower, derived from the differential distribution of blue pigments forming lines inside the tubular corolla, making recognition of species easier for the bees. In addition to the floral appealing traits such as fragrance, colour and shape, fluorescence emission of petals has been suggested as a potential signal for attracting pollinator's attention (Whitney et al. 2009; Iriel and Lagorio, 2010). Flowers of the rest of the species have patterns reflective in the ultraviolet region, which produce "nectar guides" invisible to man, which aid the bees in locating nectar or pollen resources. The vision of bees is shifted towards the shorter wavelengths of the light spectrum, so besides preferable yellow and blue they also see the flowers in UV light (Harborne, 1988). UV fluorescence may be a common trait to most flowers, but might be of temporary occurrence for parts of the flower such as stamens, style, pollen grains, nectaries and nectar (Thorp et al., 1975; Mačukanović-Jocić et al., 2007). In common with some Lamiaceae flowers, such as reported for *Ocimum basilicum*, and many other flowers (Mačukanović-Jocić et al., 2007; Iriel and Lagorio, 2010) there is a strong UV reflectance of stamens (anthers) in investigated flowers exhibiting a beautiful UV autofluorescence, which might serve as a way to catch the bee's attention. Based on our results, highly positioned anthers beneath upper corolla lip would function to facilitate foraging efficiency, whereas lower lip-positioned anthers, such as those in hyssop flower, control the landing position of visitors.

Flower morphology of two studied *Salvia* species are completely different, excluding the staminal lever mechanism, indicating that they are attractive to different insect pollinators. While *S. sclarea* flowers was only specialized on large carpenter bee, *S. officinalis* retained the ability to be pollinated by a few small bee visitors.

In general, flowers are variable among *Salvia* species indicating the adaptation to a variety of pollinators. Flower diversity in *Salvia* species due to adaptation to bee or bird pollination, especially with respect to functional lever morphology, was widely discussed by Wester and Clasen-Bockoff (2007).

Despite of being the most aromatical and rich in nectar, *S. sclarea* wasn`t at all attractive to honey bee. Unattractive flowers of *S. sclarea* are a combination of not preferable flower colour, repellent scent and morphological incompatibility which refers to the absence of co-adaptation to honey bee pollination. Although the solitary carpenter bee is polylectic, it co-evolved with *S. sclarea* and is specialized for pollinating it.

These two species have the same origin – South Europe, and orientated to each other they have developed a mutuallistic relationship providing the survival of both partners. The female of *Xylocopa violacea* was observed by Kugler (1972) as the only effective pollinator of *S. sclarea* whence, in regard to co-adaptation, it could be termed *Xylocopa*-flower (Figure 24).

Photo: M. Mačukanović-Jocić and G. Jocić.

Figure 24. *X. violacea* was observed to be the only efficient pollinator of *S. sclarea* flowers. Co-adaptation of clary sage and carpenter bee is reflected in the unmistakable functioning of staminal lever mechanism and the apropriate distance between nectar reservoir and pollen-sacs which is compatible with the proboscis length, ensuring pollen deposition on the bee's back touching the thecae while feeding. Visiting another flower of the same species permitted pollen transfer to the stigma. Pančevo, Serbia, June 2004.

Additionally, flowering phenophase of *S. sclarea* is in conformity with the *Xylocopa* life cycle. Although, according to Kugler (1972), the characteristic qualities of *Xylocopa*-flowers are also adapted to bumblebee pollination, no bumblebees have visited it during the investigation period.

Emission of odours and corolla colour are not enough to induce repeated insect visits, however, nectar and pollen are important as a food source. Phenomenon of flower attractiveness is very complex and should not be discussed separately from the pollinator behaviour related to forage. Contrary to most other bee species, the honey bees expressed different behaviour strategies in melliferous plant selection. One of the most obvious observed behaviour rules of this polylectic insect during food searching was their decision to visit no more than one plant species at the same time, despite massive simultaneous flowering of other species growing in the same habitat. For example, it was noticed that honey bees were stoutly visiting few available open lavender flowers during the day despite of great offer of other melliferous species in the vicinity. Focusing on only one species and ignoring the others is the basic prerequisite of the realisation of pollination mission. Although the pollinator's devotion is of crucial importance for plant, Gegear and Laverty (2005) pointed out that it has not been demystified what the benefit for the animal is. Animals are not aiming at pollen dispersal, but at getting the food as fast as possible. In all probability, a pollinator has to probe and learn how to approach a certain flower type along with minimal waste of time and energy. By unerring repetition of approaching procedure the most efficient ones get the tasty reward. Honey bees increase their foraging efficiency by accepting less emptied flowers and rejecting more emptied flowers than would be expected if they forage randomly. This agrees with Wetherwax (1986), who stated that honey bees express discrimination-behaviour during foraging gaining support from nectar odour or scent left by previous visitors to the flower.

As pointed out elsewhere (Dafni et al., 1988), there is no clear and precise correlation between single or combination of floral characteristics with honey bee visits. Namely, the greater nectar amount per flower and greater flower did not mean more visits as proven in *S. sclarea*. The most visited species, *H. officinalis* and *L. officinalis,* except for differences in floral productivity, had relatively small flower size with small nectar amount. Furthermore, differences in behaviour during foraging could be a result of interaction with attractive or repellent floral characteristics.

Five studied Lamiaceae species are characterized by a relatively long flowering period, ranging from two to five months, which beneficially affects the honey bee colonies at the time of their nectar and pollen collecting activities. Species that were most frequently visited, such as *L. officinalis, H. officinalis* and *S. officinalis* are potentially valuable forage plants that might be planted by beekeepers to support local honey bee populations.

ACKNOWLEDGMENTS

This study was supported by the Ministry of Education, Science and Technological Development of Serbia, Projects No 46009, as well as by EU Commission project AREA, No 316004.

REFERENCES

Abu-Asab, M. S. and Cantino, P. D. (1989). Pollen morphology of *Trichostemma* (Labiatae) and its Systematic Implications. *Syst. Bot.*, 14, 359-369.

Abu-Asab, M. S. and Cantino, P. D. (1992). Pollen morphology in subfamily Lamioideae (Labiatae) and its phylogenetic implications. In: *Advances in Labiatae Science.* (Eds.): Harley, R. M. and Reynolds, T., Royal Botanic Gardens, Kew, 97-112.

Abu-Asab, M. S. and Cantino, P. D. (1993). Phylogenetic implications of pollen morphology in tribe Ajugeae (Labiatae). *Syst. Bot.*, 18, 100-122.

Abu-Asab, M. S. and Cantino, P. D. (1994). Systematic implications of pollen morphology in subfamilies Lamioideae and Pogostemonoideae (Labiatae). *Annals of the Missouri Botanical Garden*, 81, 653-686.

Arnold, S., Le Comber, S., Chittka, L. (2009). Flower color phenology in European grassland and woodland habitats, through the eyes of pollinators. *Israel Journal of Plant Sciences*, Vol. 57, 211-230.

Baker, H. G., Baker, I. (1983) Floral nectar sugar constituents in relation to pollinator type, *117-141 from Handbook of experimental pollination biology*, (Eds.), C. E. Jones and R. J. Little. Scientific and Academic Edition. New York: Van Nostrand Reinhold Co.

Badamtsetseg, B., Sang-Myoung Lee, Hwa-Yuon Lee (2012). Pollen Morphology of the Family Lamiaceae in Mongolia. *Journal of Korean Nature*. Vol. 5. No. 2, 169-179.

Beker, R., Dafni, A., Eisikowitch, D., Ravid, U. (1989). Volatiles of two chemotypes of *Majorana syriaca* L. (Labiatae) as olfactory cues for the honeybee. *Oecologia* 79, 446-451.

Bernardello, G. (2007). A systematic survey of floral nectaries. In: *Nectaries and Nectar* (Eds.: Nicolson, S. W., Nepi, M., Pacini, E.). Springer, 19-122.

Boëtius, J. (1948). Űber den Verlauf der Nektarabsonderung einiger Blütenpflanzen. *Beih Schweiz Bienenz* 2, 257-317.

Blaženčić, Ž., Grabeljšek, T., Mačukanović, M. (1994). Apiflora of some mesophylous meadows in the Velika Morava Valley. *Acta Veterinaria*, Belgrade, Vol. 44, No. 4, 245-252.

Brantjes, N. B. M., De Vos, O. C. (2006). The explosive release of pollen in flowers of *Hyptis* (Lamiaceae). *New Phytologist,* Volume 87 (2). 425-430.

Bucher, E., Kofler, V., Vorwohl, G., Ziegler, E. (2004). Das Pollenbild der Südtiroler Honige. Landesagentur für Umwelt und Arbeitsschutz, *Biologisches*, pp. 676.

Cantino P. D. and Sanders, R. W. (1986). Subfamilial classification of Labiateae, *Systematic Botany* 11 (1), 163-185.

Carr, D. J. and Carr, S. G. M. (1987). *Eucaliptus II – The rubber cuticle and other studies of the Corymbosae Phytoglyph Press*, Canberra, Australia.

Carnat, A. P., Carnat, A., Fraisse, D., Lamaison, J. L. (1998). The aromatic and polyphenolic composition of lemon balm (*Melissa officinalis* L. subsp. *officinalis*) tea. *Pharmaceutica Acta Helvetiae* 72(5), 301-305.

Celenk, S., Dirmenci, T., Malyer, H., Bicakci, A., (2008). A palynological study of the genus Nepeta L. (Lamiaceae) (2008). *Plant Systematics and Evolution*, Volume 276, Issue 1-2, 105-123.

Claβen-Bockhoff, R. (2007). Floral Construction and Pollination Biology in the Lamiaceae. *Annals of Botany* 100(2), 359-360.

Claβen-Bockhoff, R., Speck, T., Tweraser, E. Wester, P., Thimm, S., Reith, M. (2004). The staminal lever mechanism in *Salvia* L. (Lamiaceae): a key innovation for adaptive radiation? *Organisms, Diversity and Evolution* 4, 189-205.

Comba, L., Corbet, S. A., Hunt, L., Warren, B. (1999). Flowers, Nectar and Insect Visits: Evaluating British Plant Species for Pollinator-friendly Gardens. *Annals of Botany* 83, 369-383.

Corbet, S. A. (1978). Bee visits and the nectar of *Echium vulgare* L. and *Sinapis alba* L. *Ecol. Entomol.* 3, 25-37.

Corbet, S. A., Saville, N. M., Prys-Jones, O. E., Unwin, D. M. (1995). The competition box: a graphical aid to forecasting pollinator performance. *Journal of Applied Ecology* 32, 707-719.

Corbet, S. A. (2003). Nectar sugar content: estimating standing crop and secretion rate in the field. *Apidologie*, 34, 1-10.

Cruden, R. W., Toledo, V. M. (1977). Oriole pollination of *Erythrina breviflora* (Leguminosae): Evidence for a polytopic view of ornithophily. *Plant. Syst. Evol.* 126, 393-403.

Cruden, R. W. and Hermann, S. M. (1983). Studing nectar? Some observations on the art, In: Bentley, B., Elias, T. (Eds.), *The Biology of Nectaries*, 223-242. New York: Columbia Univ. Press.

Cruden, R. W., Hermann, S. M., Peterson, S. (1983). Patterns of nectar production and plant-pollinator coevolution. In: Bentley, B. and Elias, T. (Eds.), *The biology of nectaries*, 80-125. New York: Columbia Univ. Press.

Da Silva, F. O., Viana, B. F., Jacobi, C. M. (2005). Floral biology of *Eriope blanchetii* (Lamiaceae) in coastal sand dunes of NE Brazil. *Austral Ecology*, Volume 30(3), 243-249(7).

Dajić, Z., Donnelly, R. and Helberg, U. (2003). *Balkans Herbal Development Initative - Final Summary Report - Serbia and Montenegro*, SEED and CCF.

Dajić, Z. and Kojić, M. (1999). *The distribution and representation of species in the genus Salvia in the plant communities of Serbia*, In: Brkić, D. (eds.), Sage (*Salvia officinalis* L.) – Institute for the Medicinal Plant Research "Dr Josif Pančić", Belgrade and Art Grafik, Belgrade.

Dafni, H., Lensky, Y., Fahn, A. (1988). Flower and nectar characteristics of nine species of Labiatae and their influence on honeybee visits. *J. Apic. Res.* 27(2), 103-114.

Danon, J., Blaženčić, Ž., Zonjić, T. (1990). Studies on the biological characteristics of melliferous flora providing honey-bee pasture on Goč Mountain and its surroundings, *Acta Veterinaria* 40, 77-84.

Davis, A. R., Peterson, R. L. and Shuel, R. W. (1986). Anatomy and vasculature of the floral nectaries of *Brassica napus* (Brassicaceae). *Can. J. Bot.* 64, 2508-2516.

Davis, A. R. Peterson, R. L. and Shuel, R. W. (1988). Vasculature and ultrastructure of the floral and stipular nectaries of *Vicia faba* (Leguminosae). *Can. J. Bot.* 66, 1435-1448.

Davis, A. R. and Gunning, B. E. S. (1991). The modified stomata of the floral nectary of *Vicia faba* L. 2. Stomatal number and distribution as selection criteria for breeding for high nectar sugar production. *Acta Hort.*, 288, 329-334.

Davis, A. R. and Gunning, B. E. S. (1992). The modified stomata of the floral nectary of *Vicia faba* L.1. Development, anatomy, ultrastructure. *Protoplasma* 164, 134-152.

Davis, A. R. and Gunning, B. E. S. (1993). The modified stomata of the floral nectary of *Vicia faba* L.3. Physiological aspects, insluding comparisons with foliar stomata. *Bot. Acta*, 106(3), 241-253.

Davis, A. R., Sawhney, V. K., Fowke, L. C., and Low, N. H. (1996). Floral nectar secretion and ploidy in *Brassica rapa* and *B. napus* (Brassicaceae). II Quantified variability of nectary structure and function in rapid-cycling lines. *Ann. Bot.*, 77. 223-234.

Davis, A. R. (2001). Seraching and Breeding for Structural Features of Flowers Correlated with High Nectar-Carbohydrate Production. *Acta Hort.* 561, ISHS 2001.

Delgado, G. (2000). Patterns of insect flower visitation in *Lavandula buchii* Webb (Lamiaceae), an endemic shrub of Tenerife (Canary Islands). *Journal of Natural History*, Volume 34(11), 2145-2155.

Diklić, N. (1974). Lamiaceae – *Hyssopus officinalis, Lavandula officinalis, Melissa officinalis, Salvia officinalis, Salvia sclarea.* In: *Flora of SR Serbia*, Vol. 6, (Eds.) Josifović, M., SANU, Beograd.

Durkee, L. T. (1983). The ultrastructure of floral and extrafloral nectaries. In: B. Bentley and T. Elias (Eds.), *The biology of nectaries*, 1-29. New York: Columbia University Press.

Eisikowitch, D., Masad, Y. (1980). Nectar-yielding plants during the dearth season in Israel, *Bee World* 6, 11-8.

Emboden, W. A. (1965). Pollen morphology of the genus *Salvia*, section Audibertia. *Pollen et spores* 6, 527-536.

Erdtman, G. (1945). Pollen morphology and plant taxonomy. IV Labiatae, Verbenaceae, Avicenniaceae. *Svensk. Bot. Tidskr.* 39, 279-285.

Erdtman, G. (1952). Pollen Morphology and Plant Taxonomy. *Angiosperms.* Chron. Bot Co., Waltham, Massachusettes.

Erdtman, G. (1971). *Pollen Morphology and Plant Taxonomy (Angiosperms).* Hafner publishing company, New York.

Eriksson, M. (1977). The ultrastructure of the nectary of red clover (*Trifolium pretanse*). *J. Apic. Res.* 16,184-193.

Faegri, K. *and* Van der Pijl, *L. (1979).* The principles of pollination ecology. 3rd revised edition. *Oxford: Pergamon Press.*

Faegri, K., Iversen, J. (1989). *Textbook of Pollen Analysis.* 4th ed. New York: John Wiley and Sons.

Fahn, A. (1949). Studies in the ecology of nectar secretion. *Palest. J. Bot.* (Jerusalem) Ser. 4, 207-224.

Fahn, A. (1979a). *Secretory Tissues in Plants*, Academic Press, London.

Fahn, A. (1979b). Ultrastructure of nectaries in relation to nectar secretion. *American Journal of Botany* 66, 977-985.

Fahn, A. (1988). Secretory tissues in vascular plants. *New Phytol.* 108, 229-257.

Fahn, A. (1990). *Plant anatomy.* Fourth Edition, Pergamon Press, Oxford, New York, Beijing.

Fahn, A. (2000). Structure and function of secretory cells. In: Hallahan, D. L., Gray, J. C., (Eds.) *Advances in botanical research.* Vol. 31. Plant trichomes. Academic Press, 37-75.

Fahn, A., Shimony, C. (2001). Nectary structure and ultrastructure of unisexual flowers of *Ecballium elaterium* (L.) A. Rich. Cucurbitaceae and their presumptive pollinators. *Annals of Botany*, 87, 27-33.

Farkas, Á. and Orosz-Kovács, Z. (2003). Nectar secretion dynamics of Hungarian local pear cultivars. *Plant Syst. Evol.* 238, 57-67.

Farkas, Á., Zajácz, E. (2007). Nectar Production for the Hungarian Honey Industry. *The European Journal of Plant Science and Biotechnology* 1(2), 125-151.

Freeman, C. E. and Head, K. C. (1990). Temperature and sucrose composition of floral nectars in *Ipomopsis longiflora* under field conditions. *Southwestern Nturalistt* 35, 243-426.

Fritsch, K. von (1950). "*Bees, Their Vision. Chemical Senses and Language.*" Cornell, Ithaca, New York.

Gaffal, K. P., Heimler, W. and El-Gammal, S. (1998). The floral nectary of *Digitalis purpurea* L., structure and nectar secretion. *Annals of Botany*, 81, 251-262.

Gegear, R., J. Laverty, T. M. (2005). Flower constancy in bumblebees: a test of the trait variability hypothesis. *Animal Behaviour* 69 (4), 939-949.

Gottsberger, G. T., Arnold, T. and Linskens, H. F. (1990). Variation in floral nectar amino acids with ageing of flowers, pollen contamination and flower damage, *Isr. J. Bot.* 39, 167-176.

Grabeljšek, T. 1996. The palynomorphology of melliferous plants in meadow and forest communities in north-western Šumadija. *Ph.D. thesis*, Faculty of Biology, University of Belgrade, 220 pp.

Grozeva, N. (2011). Possibilities for providing bee pasture from nectariferous plants in Sinite Kamani Natural Park – Sliven. *Trakia Journal of Sciences* 9(2), 15-21.

Gulyás, S. (1967). Zusammenhang zwischen Strukktur und Produktion in den Nektarien einiger *Lamium*-Arten. *Acta Biol. Szeged* (N.S.) 13, 1-10.

Gupta, A., Sharma, C. (1990). Polymorphism in pollen of *Salvia leucantha* (Lamiaceae). *Grana* 29, 277-284.

Hainsworth, F. R. and Wolf, L. L. (1972a). Energetics of nectar extraction in a small, high altitude, tropical hummingbird, *Selasphorus flammula. J. Comp. Physiol.* 80, 377-387.

Harborne, J. B. (1988). Biochemistry of Plant Pollination. In: *Introduction to Ecological Biochemistry*. 3rd edition, London, San Diego, New York: Elsevier Academic Press.

Harley, M. M., Paton, A., Harley R. M., Cade P. G., (1992). Pollen morphological studies in tribe Ocimeae (Nepetoidea: Labiatae) 1. *Ocimum* L. *Grana*, 31, 161-176.

Heil, M., B. Fiala, B. Baumann, and K. E. Linsenmair, (2000). Temporal, spatial and biotic variations in extrafloral nectar secretion by *Macaranga tanarius*. *Functional Ecology* 14, 749-757.

Herrera, C. M. (1988). Variation in mutualisms: the spatio-temporal mosaic of an insect pollinator assemblage. *Biological Journal of the Linnean Society* 35, 95-125.

Herrera, C. M. (1989). Pollinator abundance, morphology, and flower visitation rate: analysis of the "quantity" component in a plant-pollinator system. *Oecologia* 80, 241-248.

Houdret, J. (2002). *Herbs*. Hermes House, London, 256.

Inouye, D. W. (2004) The effect of proboscis and corolla tube lengths on patterns and rates of flower visitation by bumblebees. *Oecologia,* Volume 45(2), 197-201.

Iriel, A., Lagorio, M. G. (2010), Is the flower fluorescence relevant in biocommunication? *Naturwissenschaften* 97, 915-924.

Jakobsen, H. B. and Kristjansson, K. (1994). Influence of Temperature and Floret Age on Nectar Secretion in *Trifolium repens* L., *Annals of Botany* 74, 327-34.

Janković, M. (1982). Contribution to the knowledge of Sarplanina vegetation with a special emphasis on some important relict plant species. *Bulletin of the Institute for Botany and Botanical Garden of the University of Belgrade*. Vol. XIII, 15 (1-3), 75-129.

Jančić, R., Stošić, D., Mimica-Dukić, N., Lakušić, B. (1995). *Aromatične biljke Srbije*. Dečje novine, Gornji Milanovac, pp. 297.

Jarić, S., Mačukanović-Jocić, M., Mitrović, M., Pavlović, P. (2013). The melliferous potential of forest and meadow communities on Mount Tara (Serbia). *Environ. Entomol.* 42(4), 724-732.

Kelber, A. (2002). Pattern recognition in a hawkmoth: innate preferences, learning and ecology. *Proc. R Soc. Lond.* Ser. B 269, 2573-2577.

Kelber, A. (2003). Sugar preferences and feeding strategies in the hawkmoth *Macroglossum stellatarum*. *J. Comp. Physiol. A* 189, 661-666.

Kevan, P. G. (1983). Floral colors through the insect eye, what they are and what they mean, *3-30 from Handbook of experimental pollination biology*. (eds.), C. E. Jones and R. J. Little. Scientific and Academic Edition. New York, Van Nostrand Reinhold Co.

Kišgeci, J. (2002). *Medicinal plants: cultivation, collecting, usage* – Belgrade, Partenon, 364.

Kleber, E. (1935). Hat das Zeitgedächtnis der Bienen biologische Bedeutung? *Zeit. Vergleich. Physiol.* 22,221-262.

Kojić, M. and Gajić, M. (1973). Comparative study of ecophysiological and anatomical characteristics of some Mediterranean and sub-Mediterranean species. *Bulletin of the Faculty of Forestry, serie A – Forestry*, 44 (2), 99-116.

Kugler, H. (1972). Zur Bestaubung von *Salvia sclarea* L. durch Holzbienen (*Xylocopa violacea* L.). *Plant Systematics and Evolution*, Vol. 120(1-2), 77-85.

Lake, J. C. and Hughes, L. (1999). Nectar production and floral characteristics of *Trapaeolum majus* L. grown in ambient and elevated carbon dioxide. *Ann. Bot.* 84. 535-541.

Lazarov, A. C., Nedalkov, S. T., Mitev, B., Radoev, L., Biżev, B., and Petkov, V. (1976). *Beekeeping encyclopedia*, Nolit, Belgrade.

Leiss, K. A. and Klinkhamer, P. G. L. (2005). Genotype by environment interactions in the nectar production of *Echium viulgare*. *Functional Ecology*, 19, 454-459.

León-Arencibia, C. M. and La-Serna Ramos, I. E. (1992). Palinological study of *Lavandula* (sect. Pterostoechas Labiatae) Canario-maderiense endemics. *Grana* 31. 187-195.

Link, D. A. (1992). The floral nectaries in the Limnanthaceae, *Plant Systematics and Evolution* 179, 235-243.

Mačukanović, M., Grabeljšek, T. (1997). Apiflora dve šumske zajednice Fruške Gore. *Ekologija* (Belgrade), Vol. 33/1,2.,135-140.

Mačukanović, M., Grabeljšek, T. and Blaženčić, Ž. (1996). Apiflora of some meadows in Kragujevac surroundings (Serbia). *Acta Veterinaria* (Begrade), Vol. 46, No (5-6), 349-358.

Mačukanović, M., Blaženčić, Ž. (1998). Diurnal and seasonal dynamics of nectar secretion of some species in the family Lamiaceae, *Acta Veterinaria* (Beograd), Vol. 48, 173-184.

Mačukanović-Jocić, M., Duletić-Laušević, S., Jocić, G. (2004). Nectar production in three melliferous species of Lamiaceae in natural and experimental conditions. *Acta Veterinaria* (Belgrade). Vol. 54, No 5-6, 475-487.

Mačukanović-Jocić, M., Đurđević, L. (2005). Influence of microclimatic conditions on nectar exudation in *Glechoma hirsuta* W. K., *Archiv of Biological Sciences*, 57 (2). 119-126.

Mačukanović-Jocić, M., Rančić, D., Dajić Stevanović, Z. (2007). Floral nectaries of basil (*Ocimum basilicum*): Morphology, anatomy and possible mode of secretion, *South African Journal of Botany* 73. 636-641.

Mačukanović-Jocić, M., Stevanović, Z., Mladenović, M., and Jocić, G. (2011). Flower morphophysiology of selected Lamiaceae species in relation to pollinator attraction. *Journal of Apicultural Research* 50(2).89-101 (2011) IBRA 2011 DOI 10.3896/IBRA.1.50.2.01.

Mačukanović-Jocić, M., Dajić Stevanović, Z., Jarić, S., Đurđević, L. (2008). Nectar secretion in basil (*Ocimum basilicum* L.) grown in different soil conditions. *Journal of Apicultural Research,* Vol. 47(1). 89-90.

Mačukanović-Jocić, M. (2010). *The biology of melliferous plants with an atlas of Serbian apiflora.* Faculty of Agriculture, Belgrade 420 pp. (in Serbian).

Malahova-Akovleva, L. N. (1966). Some characteristics of flowering biology and nectar production of sunflower, *Research in Beekeeping* (In Russian).

Marden, J. H. (1984). Intrapopulation variation in nectar secretion in *Impatiens capensis*. *Oecologia* 63, 418-422.

Mc Kenna, M. A., Thomson, J. D. (1988). A technique for sampling and measuring small amounts of floral nectar. *Ecology*. 69 (4), 1306-1307.

Mitchell, R. J. (2004). Heritability of nectar traits: why we know so little? *Ecology*, 85, 1527-1533.

Moon, H. K. and Hong, S. P. (2003). Pollen morphology of the genus *Lycopus* (Lamiaceae): *Ann. Bot. Fennici* 40, 191-198.

Moon, H. K., Vinckier, S., Walker, J. B., Smets, E., Huysmans, S. (2008a). A search for phylogenetically informative pollen characters in the subtribe Salviinae (Mentheae: Lamiaceae). *Int. J. Plant Sci.* 169, 455-471.

Moon, H. K., Vinckier, S., Smets, E., Huysmans, S., (2008b): Comparative pollen morphology and ultrastructure of Mentheae subtribe Nepetinae (Lamiaceae). *Review of Palaeobotany and Palynology* 149 (2008), 174-186.

Nepi, M., Guarnieri, M., Pacini, E. (2001). Nectar secretion, reabsorption and sugar composition in male and female flowers of *Cucurbita pepo*. *International Journal of Plant sciences*, 162, 353-358.

Nepi, M., Pacini, E., Nencini, C., Collavoli, E., Franchi, G. G. (2003). Variability of nectar production and composition in *Linaria vulgaris* (L.) Mill. (Scrophulariaceae). *Plant Systematics and Evolution*, 238, 109-118.

Nepi, M. (2007). Nectary structure and ultrastructure. In: Nicolson, S. W., Nepi, M. and Pacini, E. (eds.), *Nectaries and Nectar*, 129-166. Springer.

Nicolson, S. W. and Nepi, M. (2005). Dilute nectar in dry atmospheres: nectar secretion patterns in Aloe castanea (Asphodelaceae). *International Jopurnal of Plant Sciences*, 166, 227-233.

Nicolson, S. W. (2007). Nectar consumers. Nicolson, S. W., Nepi, M., Pacini, E. (eds.), *Nectaries and Nectar*, 167-214. Springer.

Orosz-Kovács, Zs., Gulyás, S. and Halászi, Zs. (1989). Periodicity of nectar production of sour cherry cv. Pándy. *Acta Bot. Hung*. 35, 237-244.

Orosz-Kovács, Zs., Gulyás, S. and Kaposvári, F. (1992). Pollination biology of sour cherry varieties of protogyn blossoming. *Acta Biol. Szeged* 38, 47-55.

Orosz-Kovács, Zs., Nagy Tóth, E. and Mühlenkampf, E. (1994). The effect of microclimatic factors on the attractivity of apple cultivars. *XXI st Congress of the Hung. Biol. Soc., Pesc, 1994.*

Orosz-Kovács, Zs., Nyujto, F. and Kerek, M. M. (1995). The role of floral nectar production in fertility of apricots. *Acta Horticulturae* 384, 361-366.

Özler, H., Sevil Pehlivan, Ahmet Kahraman, Musa Doğan, Ferhat Celep, Birol Başer, Ahter Yavru, Safi Bagherpour (2011). Pollen morphology of the genus *Salvia* L. (Lamiaceae) in Turkey. *Flora - Morphology, Distribution, Functional Ecology of Plants,* Volume 206, Issue 4, 316-327.

Özler, H., Pehlivan, S., Celep, F., Doğan, M., Kahraman, A., Yavru Fişne, A., Başer, B., Bagherpour, S. (2013). Pollen morphology of *Hymenosphace* and *Aethiopis* sections of the genus *Salvia* (Lamiaceae) in Turkey. *Turkish Journal of Botany*. 37, 1070-1084.

Pacini, E., Nepi, M., Vesprini, J. L. (2003). Nectar biodiversity: a short review. *Plant Systematics and Evolution*, 238, 7-22.

Pacini, E., Nepi, M. (2007). Nectar production and presentation. In: Nicolson, S. W., Nepi, M., Pacini, E. (eds.), *Nectaries and Nectar*, 167-214. Dordrecht: Springer.

Pacini, E., Nicolson, S. W. (2007). Secretions analogous to nectar. In: Nicolson, S. W., Nepi, M., Pacini, E. (eds.), *Nectaries and Nectar*, 1-18. Dordrecht: Springer.

Percival, M. S. (1946). Observations on the flowering and nectar secretion of *Rubus fruticosus*, *New Phytol*. 45, 111-123.

Perišić, S., Mačukanović-Jocić, M., Karadžić, B., Djurdjević, L. (2004). The forest melliferous flora in the vicinity of Blace (Serbia). *Archives of biological sciences* 56(1-2), 39-44.

Perret, M., Chautems, A., Spichiger, R., Peixoto, M., Savolainen, V., (2001). Nectar sugar composition in relation to pollination syndromes in Sinningieae (Gesneriaceae). *Ann. Bot.* 87, 267-273.

Perveen, A., Qaiser, M. (2003). Pollen morphology of the family Labiatae from Pakistan. *Pak. J. Bot.*, 35(5), 671-693.

Petanidou, T., Van Laere, A. J. and Smets, E. (1996). Change in floral nectar components from fresh to senescent flowers of *Capparis spinosa* (Capparidaceae) a nocturnally flowering Mediterranean shrub. *Plant Syst. Evol.* 199, 79-92.

Petanidou, T., Goethals, V., Smets, E. (2000). Nectary structure of Labiateae in relation to their nectar secretion and characteristics in a Mediterranean shrub community: does flowering time matter? *Plant Systematics and Evolution*, 225, 103-118.

Petanidou, T. (2007). Ecological and evolutionary aspects of floral nectars in Mediterranean habitats. In: Nicolson, S. W., Nepi, M., Pacini, E. (eds.), *Nectaries and Nectar*, 343-375. Dordrecht: Springer.

Pozhidaev, A. (1992). The origin of three - and sixcolpate pollen grains in the Lamiaceae, *Grana* 31, 49-52.

Proctor, M., Yeo, P., Lack, A. (1996). The natural history of pollination. *London, Harper Collins,* 479 pp.

Punt, W., Hoen, P. P., Blackmore, S., Nilsson, S., Le Thomas, A. (2007). Glossary of pollen and spore terminology. *Rev. Palaeobot. Palynol.* 143, 1-81.

Rachmilevitz, T. and Fahn, A. (1973). Ultrastructure of nectaries of *Vinca rosea* L., *Vinca major* L. and *Citrus sinensis* Osbeck cv. *Valencia* and its relation to the mechanism of nectar secretion. *Ann. Bot.* 37, 1-9.

Rachmilevitz, T. and Fahn, A. (1975). The floral nectary of *Tropaeolum majus* L. The nature of the secretory cells and the manner of the nectar secretion. *Ann. Bot.* 39, 721-728.

Razem, F. A. (1998). Physiological and structural comparisons between floral - nectary and vegetative stomata in *Arabidopsis* and *Pisum. MSc. Thesis*, University of Saskatchewan, Saskatoon, Canada.

Razem, F. A. and Davis, A. R. (1999). Anatomical and ultrastructural changes of the floral nectary of *Pisum sativum* L. during flower development. *Protoplasma*, 206, 57-72.

Raw, G. R. (1963). The effects on nectar secretion of removing nectar from flowers. *Bee World* 34, 23-25.

Reith, M., Baumann, G., Claßen-Bockhoff, R., Speck, T. (2007). New insights into the functional morphology of the lever mechanism of *Salvia pratensis. Annals of Botany*. 100, 393-400.

Ricciardelli D'Albore, G. (1997). *Textbook of melissopalinology*. Apimondia Publishing House, Bucharest, Romania, 308 pp.

Ricciardelli D'Albore, G. and Intoppa, F. (2000). *Fiori e Api. La flora visitata dalle Api e dagli altri Apoidei in Europa.* Calderini Edagricole, Bologna, Italy. pp. 253.

Ricciardelli D'albore, G., Persano Oddo, L. (1981). *Flora Apistica Italiana.* Istituto Sperimentale per la Zoologia Agraria, Roma, Italy.

Robinson, G. E., Page, R. E., Strambi, C., Strambi, A. (1989). Hormonal and genetic control of behavioral integration in honey bee colonies. *Science* 246, 109-112.

Rodríguez-Riaño, T., Dafni, A. (2007). Pollen–Stigma Interference in Two Gynodioecious Species of Lamiaceae with Intermediate Individuals. *Annals of Botany* 100, 423-431.

Rusterholz, H. P., Erhardt, A., (1998). Effects of elevated CO_2 on flowering phenology and nectar production of nectar plants important for butterflies of calcareous grasslands, *Oecologia*, 113, 341-9.

Salmaki, Y., Jamyad, Z., Zarre, S., Brauchler, C. (2008). Pollen morphology of *Stachys* (Lamiaceae) in Iran and its systematic implication Flora - Morphology, Distribution, *Functional Ecology of Plants*, Volume 203, Issue 8, 627-639.

Sabsebe, D., Harley, M. (1992). Trichome, seed surface and pollen characters in *Stachys* (Lamioideae: Labiatae) in tropical Africa. In: Harley, R. M., Reynolds, T., eds. *Advances in Labiatae science.* Kew: Royal Botanic Gardens, 149-166.

Scheid-Nagy Tóth, E. (1991). Nectary structure and nectar production in apple cultivars.) *Doct. Dissertation.* JPTE, Dept. of Botany, 160.

Scheid-Nagy Tóth, E. (2000). Effect of rootstocks on primary floral attractivity in apple cultivars. *Phd dissertation.* Univ. Of Pesc, Dept. of Botany, pp. 92.

Schmidt, J. O., Johnson, B. E. (1984). Pollen feeding preference of *Apis mellifera*, a polylectic bee. *Southwestern Entomologist* **9**. 41-47.

Schultze, W., Zanglein, A., Klose, R., Kubeczka, K. H. (1989). Lemon balm, thin layer chromatography examination of the essential oil. *Dtsch. Apothek. Ztg.* **129.** 155-163.

Soursa P. (1966). Pollen morphological studies in the genus *Thymus* L. (Labiatae). *Ann. Bot. Fennici*, 3 140-146.

Stiles, F. G. (1975). Ecology, flowering phenology and hummingbird pollination of some Costa Rican *Heliconia* species. *Ecology* 56, 285-301.

Škenderov, S. and Ivanov, C. (1986). *Bee products and their use.* Nolit, Belgrade.

Szabó-Mühlenkampf, E. (1994). Nectary structure and nectar secretion of spur type apple cultivars. *Doct. Dissertation*. JPTE Dep. of Botany, pp. 159.

Thorp, R. W., Briggs, D. L., Estes, J. R., Erickson, E. H. (1975). Nectar fluorescence under ultraviolet irradiation. *Science* 189, 476-478.

Galetto, L. (1995), Nectary structure and nectar characteristics in some Bignoniaceae. *Plant Systematics and Evolution*, 196, 99-121.

Torrres, C., Galetto, L. (1998). Patterns and implications of floral nectar secretion, chemical composition, removal effects and standing crop in *Mandevilla pentlandiana* (Apocynaceae). *Botanical Journal of the Linnean Society*, 127, 207-223.

Trudel, M. C. G. and Morton, J. K. 1992. Pollen morphology and taxonomy in North American Labiatae - *Can. J. Bot*. 70, 975-995.

Tucakov, J. (1997). *Healing plants-phytotherapy Lečenje biljem-fitoterapija*. Rad, Belgrade, 687.

Umeljić, V. (1999). *Atlas of melliferous plants: In the world of flowers and bees*, I part, Kragujevac, 155.

Vasić, O. (1997). *A Survey of the Mediterranean species of Lamiaceae family in the Flora of Serbia*; Lagascalia 19(1-2). 263-270.

Vogel, S. (1983). *Nectar, Encyclopedia of Plant Physiology, New Series*, Vol. 12. C, Springer-Verlag, Berlin, Heidelberg, New York, 593-624.

Von der Oche, K. and Von der Oche, W., (2007): Celler Melissopalynologische Sammlung (CMS). *Celle's Melissopalynological Collection*, 236.

Walker, A. K., Barnes, D. K. and Furgala, B. (1974). Genetic and environmental effects on quantity and quality of alfaalfa nectar, *Crop Sci*. 1, 235-238.

Waser, N. M., Price, M. V. (1985). The effect of nectar guides on pollinator preference: experimental studies with a montane herb. *Oecologia* 67, 121-126.

Waterman, A. H. (1960). Pollen grain studies of the Labiatae of Michigan. *Webbia*, 15. 399-415.

Wester, P., Claßen-Bockhoff, R. (2007). Floral Diversity and Pollen Transfer Mechanisms in Bird-pollinated *Salvia* Species. *Annals of Botany* 100, 401-421.

Wetherwax, P. B. (1986). Why do honeybees reject certain flowers? *Oecologia* (Berlin) 69(4), 567-570.

Whitney, H. M., Kolle, M., Andrew, P., Chittka, L., Steiner, U., Blover, B. J. (2009). Floral iridescence, produced by diffractive optics, acts, as a cue for animal pollinators. *Science* 323, 130-133.

Witt, T., Jurgens, A., Geyer, R., and Gottsberger, G. (1999). Nectar dynamics and sugar composition in flowers of *Silene* and *Saponaria* species (Caryophyllaceae). *Plant Biol.* 1, 334-345.

Wolf, L. L. (1975). Energy intake and expenditures in a nectar-feeding sunbird. *Ecology* 56, 92-104.

Wyatt, R., Broyles, S. B. and Derda, G. S. (1992). Environmental influences on nectar production in. milkweeds (*Asclepias syriaca* and *A. exaltato*). *Am. J. Bot.* 79, 636-642.

Wykes, G. R. (1950). Nectar secretion researches. *Aust. Beekeeper* 52, 67-68.

Zer, H. and Fahn, A. (1992). Floral nectaries of *Rosmarinus officinalis* L. Structure, ultrastructure and nectar secretion. *Ann. Bot.* 70, 391-397.

INDEX

C

D

E

F

G

H

I

J

K

L

M

N

O

P

Q

R

S